# INDEX ON CENSORSHIP

**INDEX ON CENSORSHIP 4 2001**

INDEX ON CENSORSHIP

WEBSITE NEWS UPDATED WEEKLY

www.indexoncensorship.org
contact@indexoncensorship.org
tel: 020 7278 2313
fax: 020 7278 1878

Volume 30 No 4 October 2001 Issue 201

**Editor in chief**
Ursula Owen
**Editor**
Judith Vidal-Hall
**Web Managing Editor**
Rohan Jayasekera
**Eastern Europe Editor**
Irena Maryniak
**Editorial Production Manager**
Natasha Schmidt

**Publisher**
Henderson Mullin
**Development Manager**
Hugo Grieve
**Subscriptions Manager**
Tony Callaghan
**Project Manager**
Aron Rollin
**Director of Administration**
Kim Lassemillanté

**Volunteer Assistants**
James Badcock
Ben Carrdus
Chuffi
Verity Clark
Yasmine Gaspard
Haifa Hammami
Paul Isaacs
Andrew Kendle
Anna Lloyd
Priya Mehta
Andrew Mingay
Gill Newsham
Ruairi Patterson
Shifa Rahman
Neil Sammonds
Victoria Sams
Fabio Scarpello
Katy Sheppard
Christine Zydlo

**Cover design**
Vella Design
**Layout and production**
Jane Havell Associates
**Printed by**
Thanet Press, UK
**Previous page**
Credit: AKG London

***Index on Censorship*** (ISSN 0306-4220) is published four times a year by a non-profit-making company: Writers & Scholars International Ltd, Lancaster House, 33 Islington High Street, London N1 9LH. *Index on Censorship* is associated with Writers & Scholars Educational Trust, registered charity number 325003
***Periodicals postage***: (US subscribers only) paid at Newark, New Jersey.
Postmaster: send US address changes to *Index on Censorship* c/o Mercury Airfreight International Ltd Inc., 365 Blair Road, Avenel, NJ 07001, USA


---



---



**Cover** Jean Rustin, *'The blues have got me'*, 1998

# EDITORIAL

## Despair and destruction

In the immediate aftermath of the terrible events of 11 September, the perpetrators were regularly described as madmen. If belief in a sort of divine election, combined with indifference to the lives of the non-elect, is mad, the description is apt. In *Index*'s exploration of madness, Adam Phillips (p32) tells us that 'the history of madness is more or less a history of fear'. Certainly we have reason to fear those men flying the planes, wrapped in the cocoon of their convictions, for whom life was expendable. The thinking of suicide bombers is explained by Eyad Sarraj (p7), who for years has been dealing with the emotional fallout for the people in Gaza of the Israel/Palestine conflict. His piece makes chilling reading.

After the expressions of horror come the responses. One disturbing consequence of the attacks is the call for greater restraints on civil liberties – easier surveillance of emails, more telephone tapping, identity cards, draconian powers to hold suspects, assassination as a legitimate political tool. Historically, legislation passed immediately after a terrorist attack has tended to undermine the very rights that legislators were attempting to protect. The assassins may also have fostered support for any government with 'troublesome' Muslims – Chechens in Russia, Uigurs in China – to take action. And already there has been an alarming increase in racial abuse against Muslims in the West.

This issue of *Index* is in part about who is silenced, and why. It reminds us that, in the USSR, consigning dissenters to mental hospitals was a common form of censorship, as Vladimir Bukovsky so powerfully remembers (p92), and that silencing dissenters by defining them as mad still goes on in China (p81). Meanwhile, in Ukraine (p159), journalists are being killed for reporting on corruption in high places.

To understand what lies behind the horrific events of 11 September requires a grasp of a range of tricky issues. Among them are the Saudi connections with Muslim extremism (p14), the despair of Arab populations over American policies in the Middle East (p7), the dangers of immigrant Muslim communities choosing isolation over integration. It's vital that those who protest, argue and raise difficult questions are not demonised as troublemakers or terrorist sympathisers. It's above all a moment for us to distinguish between destruction and dissent.

**More articles, *Index* archive material and links ⇨ www.indexoncensorship.org/110901**
**Put your opinion online ⇨ www.indexoncensorship.org/comment**
**Email the editor in chief ⇨ editor@indexonline.org**

# contents

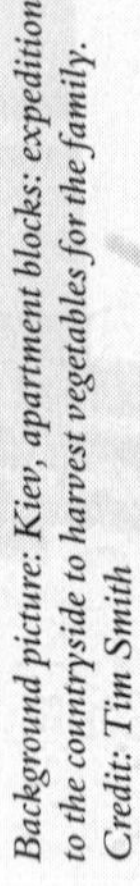

*Background picture: Kiev, apartment blocks: expedition to the countryside to harvest vegetables for the family. Credit: Tim Smith*

**Martin Rowson's 'Strip Search' p108**

# Censoring reality

When George W spoke of the 'faceless enemy' who had destroyed the 'symbol of western civilisation', he was only telling half the story: the victims of the World Trade Center remain largely faceless. Even the leading US news agency, AP, has been parsimonious in disseminating distressing images of dead and wounded; reporters on the spot tell of the site cordoned off from curious eyes: 'When they find a corpse, they remove it under cover of night.' While most European TV channels showed more than once without provoking public outrage the bodies of desperate people in the WTC falling from a great height, most US newspapers and TV channels did not; NBC, which did, was flooded with letters of outrage and reprimanded by one of its vice-presidents. Interviewed by *The New York Times*, MSNBC president Erik Sorenson confessed that he had 'cut out' any images showing 'blood and body parts'.

The country that invented the 'disaster movie', each one more inventive in its images of apocalypse than its predecessor, cannot, it seems, bear very much reality.

Voices within the US have accused it of censoring what it labelled 'dangerous' images and leaving its population with only the iconic, and in the end defiant, remains of their two towers imprinted on their consciousness in the interests of maintaining public morale and consumer spending, and preventing a stock-market crash. And, above all, to hide its collateral damage from the world. Death by terror does not have a face; only a symbol. It's a new version of the body-bag syndrome.

Tragedies happen 'out there'; not 'back here'. The bloody and dismembered bodies US channels have no qualms in flashing on prime-time TV belong to Palestinians, Israelis, Algerians, Afghans and Colombians, to name but a few. We can be certain that in any war waged 'out there' in the near future, the image will be censored and manipulated as never before. ❑

***Judith Vidal-Hall***



*Credit: AP*

EYAD SARRAJ

# Bombs and madness: understanding terror

**The creation of a 'terrorist' follows the long path of humiliation and despair**

The horrific strike against the US with a massive human life toll compels the mind to wonder how and why. How such an operation was so successful is beyond my inquiry. I shall dwell only on the 'why'.

My answers are based on my own observation as a psychiatrist living and practising in Gaza, a place I call home, which came to be known for decades as a focal point in the Palestinian scene, and the violent confrontation with Israel. Strange as it may sound, Gaza is the safest place for me, maybe precisely because it is my home. Bob Simon of *Sixty Minutes* once asked me why Gaza is safe. I replied by asking if New York was safer. Only last week, with brutal irony, Bob reminded me of that comment.

In recent years, for those who follow Middle East politics, Gaza has become notorious for producing many suicide bombers. A few years ago, I said in an interview that not only did I feel that the struggle of many Palestinians was how not to become bombers, but that the amazing thing was not the occurrence of the suicide bombing, rather its rarity.

The interviewer appeared to understand. I was shocked; it is our understanding that the world out there will never understand. And who on earth in their right mind would understand terror and the killing of innocent people? Why do Arabs kill themselves and Israelis in such a horrific way at the bus stop or in a crowded market in Tel Aviv and elsewhere?

To answer the question we must consider the social and cultural environment of Arabs and Palestinians. I believe that one of the essential elements to understand is the tribal mentality that urges individuals to avenge defeat to the bitter end, even across generations. Arabs will continue to fight for ever if need be, as long as their dignity is injured. According to the Arab tribal code, people are expected and are obliged to join the struggle. They will stop only if the aggressor publicly acknowledges his guilt and assumes responsibility for his aggression. Arabs will then enter the honourable *Solha* or peace. But only then. This code of behaviour preceded Islam and continues to be an essential expression of the identity of being an Arab.

The act of 'normal' suicide among Arabs is extremely rare and shameful, since it is perceived to be a form of defiance against God. But to be a martyr is highly glorified and places one on the highest level of respect, almost that of prophets. This is where the influence of Islam is most felt through the powerful message of the Qu'ran. Islam thus provides the vehicle for the committed to act in the deep belief that sacrificing oneself is the ultimate test of faith. More significantly, Islamic teaching is clear on martyrdom: a martyr does not die, and surely continues to live in the care of God. Muslims take that promise to heart, and literally hold to it.

Politically, suicide bombing is an act of absolute despair and a serious stage in the perception of the seemingly perpetual Arab–Israeli conflict. Citizens of any given country will join the army to defend their country. This is not so in the case of Palestinians, who have never possessed an army. They would have to join clandestine groups. Arabs also feel the despair because their regimes are not able or willing to win a fight with the formidable Israel, unconditionally supported by the US.

The road of Arabs and Palestinians to despair and to suicide killing is long. It started with the uprooting of the Palestinians in 1948 triggered by Irgun Jewish terror under the leadership of Yitzak Shamir and Menachem Begin. Since that year of the catastrophe we have tried everything. We have tried Nasser and Arab nationalism, only to be invaded in 1956 in our second homes in the refugee camps. It was only the Russian threat to bomb London and Paris, and the resolve of US President Eisenhower, that ended the Israeli occupation.

We have tried the United Nations and its Security Council which, by the way, have made excellent resolutions on our behalf even though the US has vetoed so many. For example, Resolution 194, which calls on Israel to allow us to return to our homeland, was a victory for justice.

*Palestine: Shati refugee camp, Gaza. Credit: JC Tordai / Panos Pictures*

But, like other resolutions, it was to no avail. So we kept wandering around, between airports and refugee camps, waiting for a hero or an earthquake. All we wanted was to go home.

But our story got worse and we grew bitter as we heard that a Jew from Poland would be declared a citizen of our country – now named Israel. We were told that officially we were stateless with undefined nationality. So we went to universities. We believed then that Jews were so clever because they were educated. We were told that Jews controlled the world with their education. They are doctors, lawyers and scientists, never beggars. In 20 years, many of us became university graduates; we were in every university. We had some pride. Some of our educated people formed the resistance movement. They believed that the Arab countries would never fight Israel, and that we had to force them to fight. Fatah with Yasser Arafat was born. It forced the Arabs to fight by inviting Israel to attack Egypt in 1967. Over six days, the Arabs were defeated again, but worse. This time we lost Gaza and the West Bank, Egypt lost Sinai, Syria lost the Golan. In one stroke, our fate was sealed and we had to live under Israeli military occupation for 34 years.

Do you know what it means to live under Israeli military occupation? Do you really care to know? Let me tell you a few things.

You are given an identity number and a permit to reside. If you leave the country for more that three years in succession, you lose that right to residence. Remember: this is your birthplace and is home.

When you leave the country on a trip, you are given a *laissez passer*, a travel document, valid for one year. It tells you in its recording of your particulars that you are of undefined nationality. I always found it embarrassing to fill in landing cards.

Israeli occupation means that you are called twice a year by the intelligence for routine interrogation and persuasion to work as an informer on your brothers and sisters. No one is spared. My father was asked to 'volunteer' when he was 70. I was fired from work in the children's ward in 1971 because I refused to 'cooperate'. If you are a member of a political organisation you will be sentenced to ten years. For a military action you will be sentenced to life. In both cases you will be tortured. Israel is the only country where torture was legal up to two years ago.

To survive under Israeli occupation you are given the chance to work in the jobs that Israelis do not like: sweeping the streets, building houses, collecting fruit, harvesting. You will have to leave your home in the refugee camp in Gaza at 3am, go through the road blocks and check posts, spend your day under the sun and surveillance returning home in the evening to collapse in bed for a few hours before the following day.

We simply became the slaves of our enemy. We are building their homes in our villages, and we clean their streets. Do you know what it does to you when you have to be the slave of your enemy in order to survive? No: you will never know how painful it is unless your country is occupied by another force. Only then will you learn how to watch in silence, pretending not to see the torture of your friends and the humiliation of your father. Do you know what it means for a child to see his father spat at and beaten before his eyes by an Israeli soldier?

Nobody knows what happens to our children. We don't know ourselves except that we observe that they lose respect for their fathers and identify with the new symbol of power, an Israeli soldier with his gun. So they, our children, the children of the stone as they became known, tried the Intifada – the Uprising. Seven long years our children were throwing stones and being killed daily. Nearly all our young men were arrested, the majority were tortured. All had to confess. As a result, everyone suspected that all people were spies. So we were exhausted, tormented and brutalised. Some believe that we became the mirror image of the victimised Jews.

What else could we do to return to our home? We had almost forgotten that, and all we wanted was to be left alone. What else could we try? Ah yes: peace. When the news came that Arafat had signed a peace treaty in Washington we were jubilant. At last we thought we would be rid of that miserable life of military occupation. So we had hope. At last we would be free.

We could not believe our eyes when there were no more curfews and we could actually spend our evenings on the beach or wander in streets that were now ours after 8pm for the first time in 30 years. We were ecstatic. We even had elections and a parliament, so we were told.

Suddenly, Yitzhak Rabin was murdered by a Jewish extremist. Our hopes were given a blow – also the hopes of the Israelis who started to open their eyes and hearts, at least those who accepted us as equal human beings.

Enter Binyamin Netanyahu. The new Israeli prime minister refused to meet Arafat and was clearly forced to shake hands in obvious disgust. He refused to free our prisoners, to grant a safe passage for us to move between the West Bank and Gaza. He even surrounded our towns and villages with his tanks and arrested our policemen. Then he went after our holy places and opened a tunnel under our holiest mosque.

Tens of our children, and Israeli soldiers too, were killed because of that tunnel, but he went on insulting us and driving out our sanity. Arafat called for patience and we were patient; then Netanyahu started to build settlements in Jerusalem and drive the remaining Palestinians out. Settlers in Hebron spat on our Prophet and called him a pig.

All in the name of peace, we were humiliated, arrested and tortured by Palestinian forces to protect the peace. Our Authority was trying to please Netanyahu. Our officials were driving in big cars and buying big villas. They have VIP cards and cross the check posts like human beings, while we are left to rot.

Enter Ehud Barak, another general, but claiming he would follow the footsteps of Rabin. He did not, and we were blamed. He offered to return 86% of the West Bank, only 22% of Palestine. He offered to allow 15,000 refugees out of the two million to return, and refused to acknowledge responsibility or guilt. His generous 'offer' included keeping our holy places under Israeli sovereignty.

Then enter Sharon. Oh yes: the same Sharon who was found by an Israeli commission to be responsible for the massacre of Palestinian refugees in Sabra and Shatila during his invasion of Lebanon in 1982. Sharon's election campaign started with a big bang. He forced himself into our holy mosque in Jerusalem. He wanted to tell us who the masters in this land are. He knew that we would be humiliated and angry. He knew we would throw stones. He knew his troops would kill some of us. He knew that Arafat would have no other choice but to fight. He knew that our bullets would radicalise Israel in fear. Palestinians, exhausted and emotionally traumatised, obliged, and turned their popular uprising into a deadly confrontation. Sharon, who was forced by popular demonstrations in Tel Aviv to resign because of the massacre of Sabra, won with a landslide to become prime minister. He promised peace and security. He even announced that he would be ready for *painful* concessions to the Palestinians; and he painfully described that – it was not to enter Palestinian towns!

Sharon has not put all his cards on the table; if he did, then his Zionist project of building a Jewish state on all Palestine would be in danger. His strategy is simple: provoke us to violence and show the world how evil Palestinians are as they refuse to give Israel the peace it deserves.

As our tragedy was unfolding over the last 53 years, US political leadership was determined to support Israel no matter what. President Eisenhower was the sole exception. The US supported Israel in the wars of 1948, 1967 and 1973, the invasion of Lebanon in 1982 and throughout the 34 years of its brutal occupation of the West Bank, Gaza and the Golan Heights.

We are simply baffled. When Israel occupies our land, the US asks us to negotiate! But when Iraq occupied Kuwait, it invaded Iraq. The US urges every country in the world to sign a treaty of non-proliferation of nuclear arms – except Israel. Israel never signed and possesses a nuclear arsenal. The US gives billions of dollars to Israel and many F16s, while starving Iraqi children.

As in the question of the protection of the environment from pollution, the US defies the whole world every time there is a vote condemning Israel for atrocities. US logic is that it is not good to blame Israel when the Palestinians are the party that really should be blamed.

I think I know why the US stands firm behind Israel. It has nothing to do with the interest of ordinary men and women, because the average American does not bother to find out. It has nothing to do with business, because the US pays billions of taxpayers' dollars to Israel. It has nothing to do with oil, because Arab regimes are merely humble servants of the US. It has nothing to do with fighting communism, because the Soviet Union has vanished. Nor do I believe in the theory of Islam being the new enemy, because it is not.

It has only to do with elections. The sole desire of US politicians is to be voted into Congress, the Senate or the presidency. That aim is significantly influenced by the Jewish vote, the Zionist lobby and its money. Harry Truman summed it up in the 1948 presidential election. When he was asked about the logic of not supporting the just right of the Arabs in Palestine, he replied by asking sarcastically, 'How many Arab voters are in my constituency?'

And so Israel continues to confiscate more Palestinian land, while the US pays for the building of Jewish settlements; and Arabs are humiliated by providing the cheap labour. Do you still wonder why an Arab becomes a terrorist?

The madness is like a plague. It is not confined to Arab desire for revenge, or to the Zionist dream of a pure Jewish home. It has now engulfed the US with its supra-natural array of bombs, and a Texan ride to glory. ❑

***Eyad Sarraj*** *is a physician and psychiatrist. He runs the Gaza Community Mental Health Programme*

TARIQ ALI

# Kingdom of corruption

**Keeping an eye on the ball: the Saudi connection**

The hijackers responsible for the 11 September outrage were not illiterate, bearded fanatics from the mountain villages of Afghanistan. They were all educated, highly skilled, middle-class professionals. Thirteen of the 19 men involved were citizens of Saudi Arabia. Their names are recognisable. The three Alghamdis are clearly from the Hijaz province of the Kingdom, the site of the holy cities of Mecca and Medina. Mohamed Atta, born in Egypt, travelled on a Saudi passport. Regardless of whether he gave the order or not, what is indisputable is that the bulk of Osama bin Laden's real cadres (as opposed to foot soldiers) are located in Egypt or Saudi Arabia, the two principal allies of the United States in the region except for Israel.

Support for bin Laden is strong in Saudi Arabia. He was a close friend of the Saudi boss of intelligence, Prince Turki bin Faisal al Saud, who was dismissed last month after his failure to curb attacks on US personnel in Riyadh. The real reason is probably his refusal to take sides in the fierce faction fight to determine the succession after the death of the paralysed King Fahd. Both sides are aware that too close an alignment with the US could be explosive. That is why until now the Saudi regime, despite its support for the US, is not 'allowing its bases to be used'.

In normal times, the Saudi Kingdom is barely covered by the western media. The ambassadors report to their respective chanceries that all is well and the continuity of the regime is not threatened. It requires the imprisonment of a US or UK citizen, or for a UK nurse to be chucked out of a window, for attention to focus on the regime in Riyadh. Even less is known about the state religion, which is not an everyday version of Sunni or Shia Islam, but a peculiarly virulent, ultra-puritanical strain known as Wahhabism. This is the religion of the Saudi royals, the state bureaucracy, the army and air force and, of course, Osama bin Laden, the best-known Saudi citizen in the world, currently in Afghanistan. A moderate equivalent of this in Britain would be if the Church of England was replaced by the United Reformed Church of Dr Ian Paisley, the royal family became ardent Paisleyites and the state bureaucracy and armed services were barred to non-Paisleyites.

Sheikh Mohammed Ibn Abdul Wahhab, the inspirer of this sect, was an eighteenth-century peasant who became tired of tending date palms and grazing cattle and began to preach locally, calling for a return to the pure beliefs of the seventh century. He opposed the excessive veneration of the Prophet Mohammed, denounced the worship of holy places and shrines, and stressed the 'unity of one God'. On its own this was harmless enough; but it was his social prescriptions that created problems, even in the 1740s. He insisted on Islamic punishment beatings and more: adulterers should be stoned to death, thieves amputated, criminals executed in public. Religious leaders in the region objected when he began to practise what he preached, and the local chief in Uyayna asked him to leave. Wahhab fled to Deraiya in 1744 and won over its ruler, Mohammed Ibn Saud, in 1744. Ibn Saud, the founder of the dynasty that rules Saudi Arabia today, utilised Wahhab's revivalist fervour to inculcate a sense of discipline in the tribes before hurling them into battle against the Ottoman Empire. Wahhab regarded the Sultan in Istanbul as a hypocrite who had no right to be the Caliph of Islam, and preached the virtues of a permanent *jihad* (holy war) against Islamic modernisers, hypocrites as well as the infidel. The Ottomans hit back, occupied the Hijaz and took charge of Mecca and Medina, but Wahhabi influence remained and the heroic battles became part of local folklore. This proto-nationalism was utilised by Ibn Saud's successors to expand their influence throughout the peninsula.

Two centuries later they laid the foundations of what is now Saudi Arabia, but it was the discovery of liquid gold that changed the region for ever. Fearful of competition from the UK, the US merged Esso, Texaco and Mobil to form the Arabian American Oil Company (ARAMCO). This link, established in 1933, was strengthened during World War II, when the USAF base in Dhahran was deemed crucial to 'the defense of the United States'. The Saudi monarch was paid millions of dollars to aid development in the Kingdom. The regime was a despotism, but it was seen as an important bulwark against communism and nationalism in the region and, for that reason, the US chose to ignore what took place within its borders.

The entry of the US and the creation of the Kingdom have been brilliantly depicted in one of the most remarkable contributions to Arabic fiction: the five-volume *Cities of Salt* by the exiled Saudi novelist Abdelrahman Munif, whose own birth in 1933 coincided with that of the new state. Munif's multi-layered fiction – savage, surreal and satirical – angered the royal family. He was deprived of his nationality and banned from ever returning to the country. His books became delicious contraband, circulating everywhere – including the royal palaces. When I met him about ten years ago during one of his rare trips to London, he was as lucid as ever: 'The twentieth century is almost over, but when the West looks at us all they see is oil and petro-dollars. Saudi Arabia is still without a constitution, the people are deprived of all elementary rights, even the right to support the regime without asking for permission. Women, who own a large share of private wealth in the country, are treated like third-class citizens. A woman is not allowed to leave the country without a written permit from a male relative. Such a situation produces a desperate citizenry, without a sense of dignity or belonging . . .'

Given the denial of secular openings in a society where the royal family – a clan with multiple factions and micro-factions – and its tame clerics dominate all aspects of everyday life, there were a number of rebellions in the 1960s and 70s. One of Munif's novels, *The Trench*, has a striking finale. Two revolutions are being plotted: one of them by angry young men inspired by modern ideas; the other, invisibly, inside the palace. Everything ends in tears with curfews and tanks in the street. The young revolutionaries discover that the 'wrong' revolt has succeeded. The reference was to the assassination of King Feisal in 1975 by his own nephew, Prince Faisal Ibn Musaid. Ten years earlier, Ibn Musaid's brother Prince Khalid, a fervent Wahhabite, had demonstrated in public against the entry of television into the Kingdom. Saudi police entered his house and shot him dead. To this day, Prince Khalid is venerated by hardline believers and, years later, the Taliban government paid its own tribute by publicly burning audio cassettes and videos, and banning television (*Index* 6/1998, 'All quiet in Kabul').

But Wahhabism remains the state religion of Saudi Arabia, exported with petro-dollars to fund extremism elsewhere in the world. During the war against the Soviet Union, Pakistani military intelligence requested the presence of a Saudi prince to lead the *jihad* in Afghanistan. No volunteers were forthcoming, and the Saudi leaders recommended the scion of a rich family, close to the monarchy. Osama bin Laden was despatched to the Pakistan border and arrived in time to hear President Jimmy Carter's National Security Adviser, Zbigniew Brezinski, turban on head, shout: 'Allah is on your side.'

The religious schools in Pakistan where the Taliban are still being created were funded by the Saudis; Wahhabi influence was strong. Last year, when the Taliban decided to blow up the old Buddhas, there were appeals from the ancient seminaries of Qom and al-Azhar to desist, on the grounds that Islam was tolerant. A Wahhabi delegation from the Kingdom advised the Taliban to execute the plan. They did. The Wahhabi insistence on a permanent *jihad* against all enemies, Muslim and non-Muslim, was to leave a deep mark on the young boys who later took Kabul. The attitude of the United States in those days was sympathetic. A Republican Party packed with Christian cults could hardly offer advice on this

*Afghanistan, 1996: a member of the Taliban destroys confiscated audio tape. Credit: AP*

matter, and both Clinton and Blair were keen on advertising their Christianity.

Just last year, a former liberal State Department expert on Pakistan, Stephen P Cohen, wrote in the *Wall Street Journal* (Asian edition, 23 October 2000): 'some *madrassa*, or religious schools, are excellent'. He admitted that 'others are hotbeds for *jihadi* and radical Islamic movements but these are only about 12% of the total'. These, he said, 'need to be upgraded to offer their students a modern education'.

This indulgence is an accurate reflection of the official mood before 11 September. After the collapse of the Soviet Union, the internal opposition became totally dominated by religious groups. These core Wahhabis now saw the Kingdom as degenerate because of the US connection. Others were depressed by the failure of Riyadh to defend the Palestinians. The stationing of US soldiers in the country after the Gulf War was a signal for terrorist attacks on soldiers and bases. Those who ordered these were Saudis, but Pakistani and Filipino immigrants were sometimes charged and executed in order to appease the US. The expeditionary force being despatched to Pakistan to cut off the tentacles of the Wahhabi octopus may or may not succeed, but its head is safe and sound in Saudi Arabia, guarding the oil wells and growing new arms, protected by US soldiers and the USAF base in Dhahran. Washington's failure to disengage its vital interests from the fate of the Saudi monarchy could well lead to further blow-back. They should heed the warning first sounded by the secular tenth-century Arab poet Abul Ala al-Maari, which still seems apposite:

> And where the Prince commanded, now the shriek
> Of wind is flying through the court of state:
> 'Here,' it proclaims, 'there dwelt a potentate
> Who could not hear the sobbing of the weak.' ❑

***Tariq Ali*** *is a novelist and playwright. His* The Stone Woman, *the third novel of* The Islam Quintet, *is published by Verso in paperback.* Index *6/1999, 'God is not dead', contains an important file on Saudi Arabia, available from* Index on Censorship *or online*

**Visit ⇨ www.indexoncensorship.org/saudiarabia**
**Put your opinion online ⇨ www.indexoncensorship.org/comment**
**Email the editor ⇨ editor@indexonline.org**

**PIETER-DIRK UYS**

# 'AIDS comes from Venus; HIV comes from Mars!'

Unemployed since 1974 when the South African Censor Board banned my first play. Then the second and the third. Soon it became a habit. I knew I needed to reinvent a career, if only enough to feed the cat. Laughing at fear became an instinctive therapy. The target of apartheid overshadowed everything else.

That went on for 20 years, doing my tango in front of their firing squad. Why did we in the chorus line of fascists, fools and old farts survive? Because laughter is a difficult thing to ban. And after all, here was a white Afrikaner performing jokes about his tribe and sometimes also in a dress. While that was also illegal, the legs were good.

Apartheid ended officially in 1994, and with it the long satirical road to redundancy. What more need for laughter at fear, when the fear had become farcical and politics blanded into normality? Can a minefield move and create new targets? Today I have a new role to portray: playing a middle-aged man who is terrified of dying of love. And that's a fact!

We South Africans are not sissies when it comes to viruses! The virus of apartheid lasted for over 40 years without any cure and hundreds of thousands of people died. But eventually we found a cure in democracy. And now I look for the 'mock' in democracy and try and find the 'con' in reconciliation.

South African politics was once based on total power and total fear. Fear made apartheid work. But at least you could see the virus. It was visible. It had culture, a colour. It had signs, giving its presence away: 'Whites Only.' *'Slegs Blankes.'* 'No dogs or natives allowed.' And the perpetuators of the virus lived in parliament and on the TV news.

Switch on and there's President PW Botha, licking his lips and wagging his finger! Every time we laughed at him, we felt a bit stronger and he looked a little less frightening: not less lethal . . .

The minefield has moved, from politics to sex! If politics is now just a way of life, sex is the way to death! That's really frightening. Sex is the most democratic thing in the world! Everyone is interested in Sex while not everyone is interested in politics or the vote. As was recently demonstrated in the UK! Fifty-nine per cent of the British electorate bothered to vote in the 7 June elections, while over 80% voted for *Big Brother* [popular TV show in the UK. Ed.]? George Orwell? You've got royalties due!

So the fact is: Sex kills.

The fact is: HIV leads to AIDS.

The fact is: 99.9% of the world believes that HIV leads to AIDS. However, in South Africa we would sit with that 0.1% that doesn't!

* * *

### *From the show* Foreign Aids

(*Change into the South African President Thabo Mbeki – add medical gown, surgical mask and cap. Hold pipe in hand. The delivery is slow and studied.*)

Distinguished fellow medical experts, honourable doctors, interested dissidents, members of the media:

*Sayibona.* My patients call me Dr Thaboo MacBeki . . . yes, Dr – and I know you don't believe my medical qualifications just because I'm black? Who cares. I'm in charge!

Let me start from the core of the issue and, yes, it is an interesting question and the answer is complex, so let me start off by stating most categorically that I think the most effective way to address this whole HIV/AIDS thing is to be inspired by the immortal words of Stratford-upon-Avon-born William Shakespeare, whom we now know, thanks to the African Renaissance, was the illegitimate son of a West African slave woman working in London at the time. His words sound so much more convincing in the original Swahili, but for the sake of clarity let me quote from the English translation:

To be or not to be
sure
of the link between AIDS and HIV.
That is the question
Whether 'tis nobler in the mind
to suffer the slings and arrows of the
pharmaceutical giants?
Or to take a stand against their sea of expensive drugs
and by diverting their patents destroy them?
To die? To sleep?
Perchance to dream of a Zulu witchdoctor's cure?
Aye. There's the racist rub:
for this HIV/AIDS thing is surely only but a result of
poverty and unemployment!
Two tentacles of that past fearful filth,
that octopus known by the name of apartheid.

Truly, as the great poet Shelley said, and I quote: 'I do fear it; I do fear it! Ewe I do!'

Is it possible that this AIDS will succeed where apartheid failed?

As our great Khoisan poet !X! said and I quote: '!xXxxxX'"&& !!xxxX',' rousing words eventually translated into English by Christopher Marlowe. And, thanks to the African Renaissance, we have now also discovered his Ethiopian roots. And I quote: 'Forsooth, the racist media maketh merry with the facts and addeth to out multiple confusions.'

And so it may be that I underestimate the scientific and medical expertise of those who say they know. I am ready and willing to change my views on the matter, to pay tribute to such expertise, if it is demonstrated that they do, indeed, have such expertise.

Until then, my mind is made up: don't confuse me with facts. AIDS comes from Venus and HIV comes from Mars . . .

★ ★ ★

After 1999's extraordinary experience of travelling 10,000km in a borrowed kombi with our voter-education-through-entertainment project, meeting thousands of people in townships and veld, city halls and parking lots, what was next? The 2000 municipal elections were still too vague to form the backbone for any campaign. But having ventured into a sexual minefield with my new one-man show *For Facts Sake*, a schools tour with an entertainment about AIDS/Safe Sex/Taboos/Urban Legends/etc. came to mind.

A nightmare! Where does one start? The aim would be to make the kids relax, because one was not a teacher or a parent; listen, because they'd enjoyed one's entertainment on video/TV or in the theatre. They would laugh, because there were funny things to be seen amid the darkness of HIV/AIDS. Excuse me? What's funny about AIDS! As you could ask, what was funny about apartheid? Nothing. But the fear that rules us should make us laugh, if only with relief. Once you can give fear a name it can be handled with care. Once that monster in the closet is no longer 20 metres high, but a mere few inches, it can be put into perspective. So, how to go about entertaining children with the fact that they have lost their childhood? That they are the targets of a fatal virus without understanding sex, because they're still too young to know? First: don't treat them like kids. We're all big people here, united by the terrible fact that love can kill us!

Start with a celebration of our democracy and the freedoms that have allowed all those in the school hall the right to be educated together and not separate. Remind them that there was a struggle to find that freedom, as there is a struggle to find a cure for AIDS. Picture for them recent life in a dark cloud of fear and hopelessness, which is what most of them felt for so long during those years, when white was right and black was in the kitchen . . .

Now that the arena of death has moved from politics to sex, it urgently involves those children whom life once protected with innocence. Of course innocence can be exploited as it was in the days of our childhood – the 1950s and 60s – when we were ruled by National Party politics, Dutch Reformed Church morality and that all-powerful word: ***SIES***! We knew nothing about sex, because no one spoke about it let alone answered questions. So maybe to expose one's own terror and fears in finding the way through the maze of legends and lies, of shame and sham, was an honest start? If there is one positive aspect of this plague, it is that our children will not grow up in that darkness of sexual confusion. While there are many parents who would rather tell their children about sex when 'we're ready', there is no time. When 'we' are ready they could be dead from the pitfalls of lethal sex. It's still too often '***sies***'! Often the kids are far ahead of us! So puncture the urban legends for all! Remember the 1994 election when everyone was so scared and stocked up on tins of tuna! And for what? Nothing! Like the horror urban legend that says you can get cured of AIDS if you have sex with a child? There are even taxis in Johannesburg with banners that say: 'It is illegal to have sex with a child'! This is how far the lie has become fact!

Bring out the condoms, those ugly little things that look like something dead, old spaghetti, burst balloons. Laugh and get the joke out so that when the reality happens there is no shame, no secret here.

'Girls, keep condoms with you, because boys will forget. Boys, go practise with your condoms so that you know what you're doing in the dark. You might end up by putting the condom on your big toe, because it's the biggest thing around!'

Back to girls: 'You also find out how to put on a condom, because boys are boys. They might not know, and then you sit with a virus and/or a baby. Go into the kitchen and take the broom.'

The children scream with laughter. Some can't believe their ears. Some can't believe their eyes when a banana is used in a safe-sex demonstration and then is replaced by a rubber penis 'because men and boys don't have bananas between their legs. A condom on a banana on the bedside table is not going to protect you!'

Some teachers are pale and tense. Some are relieved that it's been said, that the door has been opened. As one principal said: 'Now all we have to do to start the lesson is to say, "Remember when Uncle Pieter said that word? Well, let's talk about that."'

It's not a lecture. It's not a demonstration. It's not a talk. It's a chat among us who have to play a game of blind man's buff with all eyes open. Whereas once told: 'Don't play in the traffic, you'll get knocked down,' it is now the same warning: 'Don't play in the sexual traffic, you'll get killed!'

The bottom line is there is no safer sex than NO SEX! Sex is a precious treasure to keep. Ideally, everyone waits till they marry and then have families? Sorry, not even in the movies. Schoolkids must do homework: sex is for adults? Sorry, when nine-year-old boys are sexually active and ten-year-old girls can become mothers, who can wait with the survival kit of information? So find out all these words and names and what they mean. Don't ever be surprised by something that's been forbidden or ***SIES***!

And so, 160 schools and 300,000 kids later, a young chappie comes up for an autograph glancing casually at the banana and its pink mate on the table. Then looks up and says with a hoarse rasp: 'Jeez, Mr Uys, you're cool! No, you're Radical!' And that, as they say, is a compliment . . .

Urban legends! We in SA thrive on Urban Legends. If they didn't exist we'd make them up. Apartheid was our most successful urban legend. How anyone with a brain can think that a person of one colour is better than a person of another colour? We did for over 40 years! One urban legend that is raging around our country from township to city is the story that a man with AIDS can be cured by raping a child, raping a woman. Raping a virgin. And so, every day, ten children are being raped in Cape Town alone by frightened men who don't know what's wrong with them but have heard that this is the cure.

There is no cure for AIDS; rape is wrong! But no one's saying it! Diana is dead! She can't appear on South African Television (SATV) and say it! Thabo Mbeki says nothing! The pop idols say nothing!

Well, fuck them, someone's got to say it!

Rape is not a cure for anything!!! . . .

WWW

More articles, *Index* archive material and links ⇨ www.indexoncensorship.org/aids

Put your opinion online ⇨ www.indexoncensorship.org/comment

Email the author ⇨ pieter.d@indexonline.org

*'Evita Bezuidenhout, SA Ambassador to the Homeland Republic of Bapetikosweti'. Credit: © BME, Johannesburg*

★ ★ ★

### *From the show* Foreign Aids

(*Mrs Evita Bezuidenhout, the most famous white woman in South Africa, has been addressing the audience on South Africa and the state of the world. She talks about AIDS.*)

And I think this AIDS thing has also been pulled out of perspective. It's such a relief being here in the United Kingdom where no one's even interested. Tony Blair didn't even include AIDS in his election platform! Yes, it is a problem in South Africa and I'm very upset. It's so unfair – the poor blacks! First colonialism, then apartheid and now AIDS? But it's got nothing to do with me because I'm white! I'm a Christian! I'm Afrikaans!

Of course my children, like all young people, take it all very personally. My one son de Kock, who shares a flat with his old friend Moff de Bruyn, spends hours in front of the computer sending out emails about our AIDS orphans? And I say to him: 'De Kock! It's got nothing to do with us! It's their problem; it's their government. Once the Department of Welfare realises that they must spend their budget on the people and not themselves, the problem will be solved!'

But then my other twin son, Izan, is celebrating AIDS. My son is in jail with Eugene Terre'Blanche, the leader of the neo-Nazi right-wing Afrikaner Resistance Movement. Terre'Blanche was jailed for six years for assaulting a black man. My son volunteered to go and protect Eugene who won't eat or sleep. He just stands with his back against the cell wall 24 hours a day!

Izan says thanks to AIDS, the AWB [neo-Nazi party led by Eugene Terre'Blanche. Ed.] will one day come to power!

I said: 'What?'

'Yes,' says Izan. 'Black men refuse to wear condoms, so they will infect the black women, who will then infect the black children! So thanks to AIDS, in 20 years there will be so few blacks left that we will have a white majority government with Eugene Terre'Blanche as president.'

That I can believe: Eugene Terre'Blanche won't get AIDS; the HIV virus has got good taste!

★ ★ ★

Later I have lunch with the staff. I ask again: I didn't understand their answer. 'Do the boys have access to condoms?' It's lunchtime! You don't talk condoms at lunchtime!

'They can ask Sister.'

Sister has a moustache!

Oh, so the 15-year-old man-child who wants to fuck the 12-year-old child-man has to ask Sister for a condom?

'We can't give them condoms because it will encourage them to have sex!'

In 160 schools and prisons and colleges, not one has a condom-dispensing machine, just in case sex happens and someone needs a parachute!

The Vaal Triangle, somewhere between Johannesburg, Welkom and Hell. I think when God made the Vaal Triangle she had a migraine. Driving in my car, my plastic bag of props and a cool drink now warm on the seat next to me, in the middle of the hot Free State, flat, burnt, barren, with mine dumps and rubbish dumps and townships made of tin and cardboard?

Three schools yesterday, three today, three tomorrow. In the Vaal Triangle! I swear if I was employed to do this, I'd resign!

But then I'm in a hall. Again thousands of kids waiting to be entertained. The love affair continues!

I say to them: 'You now have the freedom to do anything you wish with your life! There's nothing to stop you from becoming the king of the castle, the chief of the *kraal*. Except tonight if you're careless? You don't get a second chance with AIDS, like we did with democracy.'

A girl stands up. Starts singing: *'Nkosi Sikelel'i Afrika'*. And they all stand up and sing, even the Afrikaans part of the anthem: *'Uit die blou van onse hemel'*. Their anthem for their future.

Afterwards, the young girl comes to me. Fourteen.

'What do I call you? Pieter? Evita?'

'Darling!'

'Tell me, is it dangerous for me to give my boyfriend a blowjob and he had got AIDS?'

Your heart stops. 'Do you kiss someone on the mouth who has a bad cold?'

'No, man, then you catch it?'

'Well, this is what can happen here too. You must find another way?'

'No, I must do it, or he beats me!'

'Well, give him a ripe paw-paw with a hole in it! Tell him to fuck the paw-paw!'

She sent me a card: 'Dear Pieter/Evita! Thanks for the good advice. It works. I no longer give him a blowjob. I gave him the paw-paw. But now he has fallen in love with the paw-paw!'

Let me show you some snapshots of my last year:

A party on Freedom Day in Darling where I live. We were in the coloured township which is called Smartietown. We were all celebrating seven years of democracy since that magic day on 27 April 1994. And we remembered how we could party on that night for the first time as a nation and not as legal whites and illegal blacks. How we danced and sang and drank and loved. And the young woman next to me lifted her glass.

'Here's to the new South Africa. On that 27 April 1994 I got the two things that changed my life: I got the vote and I got the virus!'

The phone rings late at night.

'Hello? Pieter?' I think who is this idiot doing an Evita in my ear?

'Pieter? Please come and help us. All our boys are HIV.'

A reformatory two hours from Cape Town. 150 boys, inside for murder and rape; between the ages of ten and 16. All HIV. In fact, a boy had been caught stealing a car on Friday. They found then he'd borrowed his father's car without asking. It was the weekend. He had to wait till Monday to walk free. By then he'd been raped eight times. Now he also is HIV.

What do you say to 150 kids in for rape and murder and soon AIDS as well? 'There was a man who was in jail for 27 years but he didn't give up; he came out and changed our lives'? Waiting, hoping, wanting laughing, living!

I ask: do you have access to condoms?

At the Hout Bay High School overlooking a magnificent bay and Chapman's Peak Drive, set in the coloured township where the mandrax merchants rule and the fisherfolk live, I see a girl in the third row. She looks like a *Vogue* model. She's not a model. She has AIDS. Her name is Christine.

I talk to her afterwards.

'Are you getting medical help?'

'Yes,' she says, 'I get a free pill, but the medicine is so poisonous I think I would rather die of AIDS . . .'

'Is there nothing better you could take?'

'Yes,' she smiles, 'there's a perfect cocktail of drugs, but that costs R50 a month.' R50 is £5.

But we don't have the money to give medicine to the sick. Thabo Mbeki won't allow the drug AZT to be given to pregnant mothers with AIDS to prevent their babies being born with HIV; or to raped women. (Although he does make AZT freely available to members of parliament and his ANC Politburo!)

Mbeki must also find £5 billion to pay the British government for the arms we have recently purchased! British guns with French bullets that don't fit and an instruction book in German that no one can understand? So I think Christine from Hout Bay must be patient: priority before sentiment. Have a nice day.

In the past the South African government killed people; now we just let them die! . . .

That's a new one: quality of death? Nkosi Johnson dazzled the world with his courage and honesty. This giant of a little boy was born HIV – he got it from the angels – and lived with it and died of AIDS at 12.

In five years South Africa will have two million Nkosi Johnsons! Where do they live? How do they die? On a cement floor because everyone is so scared of touching them, because our president and his government are in denial?

And it is the schoolchildren who ask: what can we do?

Look after yourself! No one will help you!

You must empower yourself with knowledge! Understand what it is. Don't be frightened. Ask questions. Talk. Protect yourself.

And find out who is sick, who's got AIDS. Don't throw stones at them or hurt them. Hug them and hold them and love them. You won't catch it.

It's easier to catch racism than it is to catch AIDS. ❑

***Pieter-Dirk Uys** is South Africa's leading satririst and performer.* Foreign Aids *opened at the Tricycle Theatre, London, in July 2001*

# The silence of madness

**Can we be sure that those we classify as mentally ill can make their voices heard? Who controls the classifiers? Do we listen to those we call mad?**

**ADAM PHILLIPS**

# Round and about madness

**'A chain of associations is to him what a chain of reasoning is to other men; and what he calls his opinions are in fact merely his tastes' – 'Southey's Colloquies', *Edinburgh Review* 50 January 1830**

When the British psychoanalyst John Rickman remarked that madness is when you can't find anyone who can stand you, he was asking a couple of questions. First, what makes us feel that we can't stand someone? And this becomes, for the sake of diagnosis as it were, a question about how we know, about how we describe what it is about them that we can't stand? And secondly, what do we tend to do when we can't stand someone? Answers to the second question constitute what we now call the history of madness, which is more or less a history of fear, at least in the modern era. A history of forms of classification; and of how the so-called mad create an unease that cannot be ignored, and for which they have often been punished by the state.

The hatred of this unease has sometimes needed explaining and hatred itself always makes us fearful – but the mad have traditionally been those people we have to do something about. It has been assumed that, as with criminals, we must do something with them or they will do something with us. Unlike criminals, they are not always people who have committed crimes. They are either felt to be harmlessly strange – withdrawn, out of reach, out of touch or often potential criminals, capable, perhaps, of crimes we have never even dreamed of. But in so far as they don't break the law – it's not, for example, against the law (yet) to hear voices or not to speak – they are more like people with disturbingly bad manners. People who by not playing the game make us wonder what the game is. And, indeed, why the rest of us have consented to play it. So the mad have also been available to idealise as cultural outlaws or odd prophets, as though madness was a glamorous misery and not a monotonous one. As though to be treated as an oracle was not itself a form of scapegoating. What we call madness, or even what we call pornography, is that to which we cannot remain indifferent. It is, in other words, something about which every one has, or takes, a position.

So there is a certain relief (along with terror) when the mad commit crimes, or when the criminal is described as mentally ill, because then their behaviour seems intelligible, straightforwardly transgressive rather than horribly eccentric. The law (like medicine) is always trying to keep up with the mad, keep track of them – and modern political regimes (like modern psychiatric approaches) have shown just how easy it is to recruit the language of mental health for diverse political objectives – because those people seem to be unable or unwilling to follow rules. Their language, their beliefs, their bodily gestures, their hygiene, their hopes and expectations can be wildly at odds with some putative norm. They remind us of what it is to be normal. As Rickman's comment suggests, what is called madness is all to do with sociability, and what sociability is all to do with.

Whether it is bandied about by anti-psychiatrists or by British democrats, the word madness covers what would once have been called a multitude of sins. But at its most minimal it is always pointing out the same thing: something so unacceptable about a person (or a group of people) – either to themselves or to others or to both – that intervention seems to be required. So-called madness can be described as a message, always enigmatic and always disturbing, that creates a certain kind of environment around itself (a world, say, of fearful, dismayed, punitive people; a world of curious, protective, kind people, and so on). It is defined, so to speak, by what it elicits in others. It feels like a call to action; at one end of the spectrum there is the silencing and segregation of the mad, at the other end there is the treatment of the mad that, at least in intention, is more sympathetically disposed. Madness may be treated as a call to action; but this action, even at its most well (or ill) intentioned, cannot simply or solely be described as relief from suffering: because to relieve someone of their suffering can be to deprive them of their point of view. To think of suffering as the worst thing we suffer from can itself be a form of censorship. There is the suffering of the supposedly mad, and the suffering caused by the mad; and the pervasive difficulty of working out who is doing what and to whom. Or rather, of working out which kinds of conflict we can bear to value. This, as Rickman intimated, is to do with the consensus in any given society about what are considered to be the valuable pleasures.

The word 'mad' is normally used about someone when his or her sociability begins to break down, but in a way that seems to endanger sociability itself: no one understanding what someone is saying, everyone feeling intimidated by someone. Just what it is that madness seems to endanger – how these forms of classification, these rhetorical terms and figures operate in a given society – is what the mental health professions have to account for in order to justify their existence. Talking about madness, in other words, is a way of talking about our preferred versions of a life; of what it is about ourselves and our societies that we want to protect and nurture (if we can), and what it is about ourselves and our societies we would prefer to be rid of (if we can). So the political implications of the concept and the category of madness – the consequences of speaking the diagnostic languages of mental health, rather than more simply saying who we are prepared to listen to and why – seem self-evident. And yet there are always going to be people who feel more comforted and consoled – more understood, ie, part of the group – by being diagnosed, by being subject to apparently authoritative descriptions. It is always a mixed blessing to resist the solace of sociability. If diagnosis is the problem and not the solution, if it is merely, as it were, legitimated ganging up, then those who are to resist it will need powerful counter-descriptions of their own. The question is always going to be: from where can they get the words they need, from whom can they get the emotional buoyancy that will free them to speak? It is about whether people can make any kind of group out of whatever it is that isolates them. All political regimes attempt to isolate whatever threatens them. So it always looks like a conundrum: are people mad because they are isolated, or are they isolated because they're mad? Are people different because they are oppressed or are they oppressed because they are different? The answer is clearly both, but it is the momentum of this vicious cycle that is horrifying for all concerned.

Every culture has its own ways of taking people seriously and of telling what is serious about people. And western cultures in the modern era have had institutions to train people in the forms of recognition the culture wants to promote. So there are people who know what art is, people who know what illness is, people who can decide what it is to break the law, and so on. And yet the category of madness has always been a problem wherever legitimisation has been an issue. Around and about the enigmas of legitimisation – who legitimates the legitimators; which people, or how many people, have to agree to make us believe that something is good, or true, or maybe beautiful? – what has been called madness has been whatever has put legitimisation procedures into doubt. In The Bacchae, as in King Lear, a mockery is made of what people most believe in. Madness is that which disrupts due process, as defined by the powers that be; and exposes the fact that the powers that be are not the only powers that be. It becomes the category – of thought, of feeling, of desire, of drama – that none of the categories can contain. And it crosses the disciplines, so to speak, by blurring their concerns. Whether madness is a legal issue or a medical issue; whether it requires a scientist or an artist to make it intelligible; whether indeed the artist and the scientist (and the serial killer) in some sense need to be mad to be who they are and what they do – all this makes the concept of madness akin to the universal acid of (scientific) folklore. That there could be something in a culture – call it a force, or a voice, or an energy, or a figure – that nothing and no one in the culture can command. And that it might always be baffling, whether this something was a source of renewal or a source of ruin, so that no one would ever be able to tell whether this was a good or an evil thing, a thing we should explore. These are the once-theological preoccupations that the secular notion of madness keeps in circulation. Madness becomes a way of talking about what is wrong with ways we describe what is wrong with us. It reminds us, in a technological age – rather like the body, or the notion of the unconscious – that the fact that we can describe ourselves doesn't mean that we invented ourselves. That there is more to us than we have been able to account for, and that this more may be precisely what we should pay attention to. Though not, perhaps, in the way people previously paid attention to God. When people are being excessive, we call them mad.

And yet what all the exhilarating talk (and writing) about so-called madness tends to discount – madness as demonic and chthonic; madness as parody and inspired critique – is the misery of it. And secondarily the very real difficulties – difficulties of terror and of comprehension – of being, in any capacity, with very disturbing people. Glamorising the mad may at least have encouraged people to consider the possibility of listening to people they would rather flee from; but it also idealised certain forms of suffering in a way that was sometimes unrecognisable to the sufferers. Treating people as special is often a way of neglecting them. And this is where politicising the mad and describing political dissidents in the language of mental health is pernicious. The mad are like the politically oppressed in so far as they can be the victims of a powerful consensus that needs to invalidate them; the politically oppressed are like the mad in so far as they are punished for their beliefs (their actions and their sentences). The mad are unlike the politically oppressed in so far as they may be destructive in ways they themselves can't bear: and which isolate them in ways they can't bear. The politically oppressed are unlike the mad in so far as what makes them political is that they share their beliefs with at least some other like-minded people. In short, the politically oppressed are always a group; the mad often aren't; at least from their own point of view. As Rickman intimated, you aren't mad, or you aren't that mad, if there are people who can stand you. In political groups there are often people who admire and believe in each other. Clearly the vilest political regimes destroy the opposition by most cruelly isolating them; either from the outside world, or from each other, or from both.

If the mad are traditionally those people in the culture who make inadmissible connections – who live as if they are like Napoleon, who link life with excess (of strangeness, of silence, of coherence and incoherence) – they are defined by their capacities for recruitment. When they organise certain groups of people against themselves – in our culture, psychiatrists and sometimes the police – they are called mad; when they organise people around them, speaking on their behalf, as it were, they are called religious (or cult) leaders, politicians, celebrities or very (and variously) talented people. People are not called mad when sufficiently influential people agree with what they are saying. Once one takes seriously the extent to which consensus is sovereign in matters of so-called mental health, as in so much else, one realises just how endangered any individual or group of people is once they put themselves out of range of the available culturally sanctioned descriptions. At its worst the mad, like the politically dissident, are prevented from adding to the stock of available reality. And it has, inevitably, been the official languages for managing the mad that have more or less decided where and how they can participate in the lives of those deemed to be full of mental health.

If for the sake of discussion – but not necessarily for the sake of justice – one blames neither the psychiatrists nor the mad for what they have done with each other, it is possible to see that, broadly speaking, there are two modern accounts of madness. In one account, which the big money is on, what we call madness is an organic (neurological, genetic) dysfunction of the organism. The mad are chronically maladapted bodies: they need a combination of rewiring and new chemistry to regulate, if not to cure, the destructiveness of their natures. At the extreme end of this spectrum they were born with this madness – call it schizophrenia, call it bi-polar disorder, call it depression – in their bodies. At the less extreme end something was done to their relatively normal bodies – call it a traumatic experience, a war, an illness – and it deranged the physics and chemistry of who they were. Science, which by definition understands such things, has a lot to say about this bit of the physical universe. From a scientific point of view these things are thought of as disorders rather than complaints; though this, in and of itself, doesn't make a scientific approach less compassionate. But it does assume some kind of knowledge, some kind of picture of what it is for an organism to be well-functioning. And the mad, whatever else they can do, are not good at functioning. Or at least at functioning as previously defined. The first Christians, like the first trade unionists, were obviously proposing new ways in which people might function. But of course, from the scientific psychiatrists' point of view, the person diagnosed as schizophrenic is not inventing a new way for people to live, he is unable (or failing) to live in the normal way. Indeed he needs to get back to that putatively normal state so that he is in a position to choose whether he wants to be a Christian, or to join a trade union. Mad bodies are ill equipped, or just not ready for human culture. In this view, at its most extreme the mad are not makers of meaning; they don't have the equipment for it. So, a bit like children, they are the beneficiaries and the victims of other people's meanings. They have descriptions foisted upon them from a (consensually agreed) more privileged position. Clearly, what the mad are deemed to be like – animals, children, machines which have broken down, eclipsed geniuses – will dictate how they are to be treated.

In the other modern account, madness is described as either a person's ingenious, though debilitating, self-cure for the obstacles thrown up by their individual development. And as all development is deemed to be traumatic, these mad solutions will turn up, to a greater or lesser extent, in everyone's life. Or madness is taken to be just a description of what human beings really are. In the depths, in our hearts, as passionate creatures, we are mad: in excess of the cultures we create, and always beyond our most searching descriptions of ourselves. In this view, who we are is a mockery of what we make, because who we are is fundamentally uncontainable (all our cultural forms are just ways of getting away with something). The notion that we are truly and deeply mad is, of course, far older than the notion that, given a chance, we are eminently sane. But in these accounts it is our nature to be mad, and it is our nature to protect ourselves from this madness (which psychoanalysts, for example, call desire or instinctual life). And the ways we have found to protect ourselves or cure ourselves – called defences, or symptoms, or eccentricities – are themselves mad. Madness, in short, is considered to be both the problem and the solution. So the sponsors of this account are always, one way and another, trying to make a case for some kinds of madness being better than others; the madness of sexual love, for example, is preferred to the madness of agoraphobia, or obsessive compulsive working habits and so on. But whichever madnesses are given credit, they are all assumed to be either utterly meaningful, or, indeed, to be the very source of meaning, as though what we call madness is the matrix of all our sense-making. And what we call reason is just part of our madness, one of the things our madness has got us into.

Madness as source, and madness as dysfunction are unpromising options. And it has become reasonable these days, when it comes to what was once called madness, to prefer more complex descriptions of whatever it is about other people that disturbs us, and them. Ideally, these descriptions combine rather than simply exclude the available methods and knowledge. And yet, of course, being reasonable about madnesses is at once all we can ever be – the most radical anti-psychiatrist, like the most radical biological psychiatrist must, by definition, have their own shareable logics, their own persuasively cogent arguments – and the most implausible thing we can be . All writing about madness is either proud or embarrassed about its distance from what it describes. Whether one writes deliriously or dispassionately about madness one cannot escape a binding paradox: if it can be represented (described) – in psychiatric diagnosis, in political character assassination, in lyrical description – why call it madness, rather than strange behaviour? Why not attempt, without prior categorisation, increasingly nuanced and subtle descriptions of unusual ways of being a person? Instead of bestiaries we might just have collections of tales.

The reason this may not happen is that, in actuality, madness is the word we use to refer to the violence we most fear; and so, by implication, to talk of madness is to talk, however obliquely, of whatever it is in ourselves and our societies that we dread being violated. It is the violence – real and imagined – that the so-called mad do to us, or to something about us, that is likely to make us cruel in return. It is both the drawback and the advantage of the scientific approach to madness that it starts from the position that we (the sane, the doctors) can transform them (the mad, the nominally ill) but they must not transform us. And yet, from the other point of view I sketched above – call it the bohemian approach – wherever the nature of an exchange has been decided in advance there is something called oppression at work. This is why the scientists are often more politically co-optable than the bohemians by the more repressive regimes.

It would be nicer and better if we stopped thinking in terms of reason and unreason, stopped working out who is mad and who is not, and started working out who we are prepared to listen to, and why; and, of course, what we imagine the consequences of such listening might be. But we would have to do this mindful of the very real terrors involved, and so be able to distinguish, as far as is humanly possible, between punishing people and looking after them (doing either in the name of the other is a mystification). That the language of so-called mental health experts has been so easily recruitable for nefarious ends should make us wonder whether these really are the best descriptions available for the most extreme forms of modern suffering.

If you can't find anyone who can stand you, you can't find anyone who believes you've got anything they want. Groups consist of people who, for better and for worse, need each other's company. What we call madness highlights, as it were, our infinite anxieties about exchange with other people. The anxiety of influence is as nothing compared with the anxiety of exchange. ❑

***Adam Phillips*** *is a psychiatrist and author. His latest book is* Houdini's Box *(2001, Faber & Faber)*

RENATA SALECL

# After the war is over

**The breakdown of soldiers in combat has been recognised since World War I when it was first described as 'shell shock'. The nature of modern warfare calls for radical reassessment of the phenomenon – and dangerous new solutions**

There was a little girl on the ground in purple rubber boots, not breathing. A crowd was forming around her, which was what drew Darryl King's attention in the first place. The corporal was on routine duty, closing up his peacekeeping tour in Kosovo, with Canadian Forces 15 Combat Engineers, when he saw civilians running towards a car that had stopped near his position in Priština. It was early December 1999.

What happened over the next few minutes eventually strained his family relationships, hindered his military career and, to this day, haunts his thoughts. The little girl, who had been struck by a speeding car, was actually known to King. She used to tease him when he did his daily exercise run in Priština.

But that day, he scooped her up in his arms and ran for her life towards a nearby medical unit. Later, with her blood still on his hands, King found out the child had died. He worried then – and worries now – that his spur-of-the-moment rescue run might somehow have contributed to her death. 'She was about seven or eight years old,' he says quietly. 'Her name was Hannah. She was born in war and survived it – for what? To get run over by a car?'

The incident was central to King's mental anguish on his return to Edmonton Garrison, and to his eventual diagnosis of Post-Traumatic Stress Disorder, the bane of the modern peacekeeping soldier. Those succumbing to PTSD face a classic military Catch 22: seek help and perhaps the nightmares, the chronic anxiety and the unchecked anger

*Sarajevo, Bosnia, 1998: victim of war in City Mental Hospital. Credit: Robert King / Camera Press*

will be treated, but run the risk of arousing the suspicion of fellow soldiers that you are shirking your duty.

There are numerous reports dealing not only with the suffering of the victims of the most recent wars in the Balkans, but also of those who tried to stop those wars – the peacekeepers. And the debate on their suffering seems to continue the decades-long discussion on post-war trauma. In World War I, psychologists spoke about shell shock, seeking to link psychological breakdown to the horror of shelling. In the Vietnam War, however, the concept of 'post-traumatic stress disorder' was introduced in an effort to remove guilt from the individual soldier: his condition is the result of external circumstances, not of his own psychological predisposition.

*Vietnam, 1992: ex-soldier at the Centre for Mental Treatment near Hanoi. Credit: Panos Pictures*

For decades, military psychiatrists tried to figure out what causes breakdown in the midst of combat; and why many veterans continue for years to have nightmares, depression or panic attacks, often leading to suicide. A soldier is usually able to carry on in combat for considerable time, under conditions of extreme danger and discomfort, until something happens that his defence mechanism cannot encompass.

As one psychiatrist explains: 'The actual events varied tremendously, ranging from things as simple as a friendly gesture from the enemy or an unexpected change in orders to the death of a leader or a buddy.' In all these cases, the soldiers suddenly changed their perception of the war and were unable to continue participating in the battle. These soldiers did not suddenly become cowards; rather, they experienced an acute state of anxiety, radically different from fear.

In general, we fear something we can see or hear, a tangible object or situation. Fear can be articulated: we can say, for example, 'I fear the dark,' or, 'I fear barking dogs.' In contrast, we often perceive anxiety as a state of fear that is objectless: we cannot say what it is that is making us anxious. Anxiety has a more powerful effect/affect than fear precisely because we cannot pin down the cause.

This definition of the difference between anxiety and fear corresponds to what we *think* we experience in our daily lives. Psychoanalysis, however, gives a more complex view of this difference. Freud, for instance, had two theories of anxiety: one that links it to an excess of libidinal energy that has not been discharged; another which takes anxiety as a feeling of an imminent danger to the ego, thus pushing the subject into a state of panic.

When French psychoanalyst Jacques Lacan speaks about anxiety, he introduces the problem of castration and the subject's relation with what he calls the Other – the term used to designate the social-symbolic structure in which we live: language, culture, institutions. However, it is not that the subject has some kind of a castration anxiety in regard to the Other, that he or she takes the Other to be someone who might take something precious from him or her. Lacan points out that the neurotic does not retreat from the castrating Other, but from making of his own castration what is lacking in the Other. What does this mean? When psychoanalysis claims that the subject undergoes symbolic castration by becoming a speaking being it is asserting that language marks the subject in a specific way, which makes the human being very different from other living beings. Language introduces a constutive lack into the subject, because of which, the subject is able to desire, and constantly searches for objects of this desire that might fill the void or sense of lack and make it disappear.

However, the subject not only always fails in this attempt, the subject also faces the fact that he or she has no full identity which would grant him or her some stability and consistency. When we say that the subject is castrated, we mean primarily that the subject *per se* is empty, nothing by him/herself. All the subject's power comes from the symbolic insignia he/she temporarily takes on. Take, for instance, a policeman; a nobody, perhaps, until he puts on his uniform and becomes a person with power. The subject is castrated, that is to say, powerless by himself. Only by occupying a certain place in the symbolic order does he or she temporarily acquire some power or status.

On the one hand, the subject is concerned with his or her own inconsistency (the lack that marks him or her); on the other hand, the subject is also bothered by the fact that the society in which he or she lives is marked by antagonisms, that is to say, that the Other (the symbolic order) is inconsistent – divided, non-whole. The subject can thus never get a full answer to what it is the Other wants or how the Other sees him or her. This creates a specific discomfort.

One of the ways the individual manages this unease is by creating a fantasy. Fantasy is a way of covering up the fundamental lack by creating a scenario, a story that gives his or her life a sense of consistency and stability, while he also perceives the social order as being coherent and not marked by antagonisms. Fantasy and anxiety present two different ways for the subject to deal with the lack that marks him as well as the Other, the symbolic order. If fantasy provides a certain comfort to the subject, anxiety incites the feeling of discomfort. However, anxiety does not simply have a paralysing effect. The power of anxiety is that it creates a state of preparedness so that the subject is less paralysed, less surprised by events that might radically shatter his or her fantasy and thus cause a breakdown or the emergence of trauma.

Fantasy, however, also helps prevent the emergence of anxiety. Take the case of the Israeli soldier Ami, as recorded in Zahava Solomon's *Combat Stress Reaction: The Enduring Toll of War.* Ami served in the Yom Kippur and Lebanon wars. He had been an avid filmgoer in his youth and during the Yom Kippur war he felt as if he was playing the part of a soldier in a war movie. This fantasy sustained him throughout the war: 'I said to myself, it is not so terrible. It's like a war movie. They're actors, and I'm just some soldier. I don't have an important role. Naturally, there are all the weapons that are in a war movie. All sorts of helicopters, all sorts of tanks, and there's shooting . . . [But] basically, I felt that I wasn't there. That is, all I had to do was finish the filming and go home.' Later, in the Lebanon war, Ami felt like a tourist observing pretty villages, mountains, women, etc.

But at some point, the fantasy of being on a tour or in a movie collapsed. This happened when Ami witnessed massive destruction in the Lebanon War and was involved in heavy face-to-face fighting. The scene that triggered his breakdown happened in Beirut when he saw stables piled with corpses of Arabian racehorses mingled with corpses of people. The sight filled him with a sense of apocalyptic destruction, and he collapsed: 'I went into a state of apathy, and I was not functioning.' Ami explains the process as follows: 'In the Yom Kippur war, I put my defence mechanism into operation and it worked fantastically. I was able to push a button and start it up . . . In Lebanon, the picture was clearer. In the Yom Kippur war, we didn't fight face-to-face or shoot from a short distance . . . If I saw a corpse, it was a corpse in the field. But here [in Lebanon] everything was right next to me . . . And of all things, the thing with the horses broke me . . . A pile of corpses . . . and you see them along with people who were killed. That's a picture I'd never seen in any movie . . . I began to sense the reality [that] it's not a movie any more.'

Anxiety emerges when, at the place of the lack – the point of inconsistency that the fantasy covers up – the individual encounters a certain object that perturbs his fantasy framework. For the soldier Ami, this happened when he saw the pile of dead horses. His fantasy had enabled him to observe dead soldiers because he was an outsider watching a movie; horses were something else. The fantasy collapsed and Ami broke down.

Today, there are similar problems with peacekeepers. Wendy Holden points out in *Shell Shock: The Psychological Impact of War* that peacekeepers suffer from the fact that they must observe atrocities but are helpless to fight back or defend those they have been sent to save. 'Proud to become professional soldiers and keen to fight a war, they are, however, distanced from death and the reality of killing. They are members of a society that finds fatalities unimaginable. When presented with the unimaginable, they crack.' British peacekeeper Gary Bohanna came to Bosnia with a belief that a peacekeeping role was better than a war in which colleagues got killed.

But he was quickly disillusioned when he saw civilians killed, women raped and whole families slaughtered. The traumatic event which precipitated his breakdown was when he saw a young girl who had 'shrapnel wounds in her head, half her head was blown away. Her eye was coming out of its socket and she was screaming. She was going to die, but I couldn't bear her pain. I put a blanket over her head and shot her in her head. That was all I could do.' Here again is the soldier who comes to war with the protective shield of a fantasy – this time that he is coming to do good deeds but is not fully engaged in war – only to have his fantasy undermined by events.

Unlike the recent past, in which wars still involved at least minimal engagement between the soldier and the victims on the battlefield, the soldier today is often a distant actor who shoots from afar and has no knowledge of the consequences on the front. Contemporary wars are supposed to be aseptic, so that US soldiers can fly for a couple of hours to drop bombs over Kosovo, but return home in time for football on TV. For those soldiers who will still need to engage in direct battle, military psychiatrists anticipate anxiety will be so overwhelming, so paralysing they will need a magic potion to alleviate anxiety and turn soldiers into robots with no emotional engagement with the atrocities they are committing. So far, all attempts to create such drugs have failed. The anti-anxiety drugs allegedly used on the front at the time of the Gulf War not only failed to alleviate anxiety, they produced numerous side effects that made soldiers zombie-like creatures, barely able to function or perform their duties.

The trend towards making war anxiety-free paradoxically goes hand in hand with today's attempts to make wars independent from political struggles. In the West's assessment of the situation in former Yugoslavia, the political dimensions of the conflicts were constantly overlooked or too quickly historicised. The problem with NATO interventions is that they were publicly presented as simple humanitarian missions that had nothing to do with the political situation in the region and never admitted the West's strong economic interests in the Balkans. This humanitarian ideology goes hand in hand with the media's depiction of the war. On the one hand, we are shown war as a simple computer game in which the soldiers dropping the bombs from the air are completely detached from the reality of the situation on the ground. On the other, we see images of suffering victims – the destruction of villages, people being killed, wounded and dead bodies exposed on the screen, and so on. This over-visibility of one side, and complete invisibility of the other, are linked to the fact that the economic and political logic of intervention in the war zones remained unravelled.

Advances in science and the vast cyberworld of new technologies foster the impression that life is like a computer game; that we live in a world of simulacra in which bodies and identities are something we can change and play with. The boundary between life and death also seems to be changed. In cyberspace, one can enjoy playing with numerous imaginary personas; in the near future, it will be possible to create new body parts out of existing genetic material; easy, eventually, to imagine the laboratory production of the entire human body. Given this scientific background, we can assess breakdowns in war rather differently. Freud pointed out that war forces people to deal with the question of death in a new way: they usually avoid thinking about the possibility of their own death and it even seems as if the unconscious has no image of it at all.

Today, with the proper genetic code and the invention of new drugs, people hope that matters of life and death will be more predictable and controllable in the future. In the same way, military psychiatry hopes it will be able to master the soldier's traumas in regard to death with the help of new drugs. The ideal soldier of the future will be completely detached from the situation (an outsider not really present in the war) and will neutrally observe the atrocities going on in the war. However, the problem with the military's attempts to find a cure for anxiety, for example with the help of drugs, is that rather than preventing the soldiers' anxieties, such drugs actually help to provoke new ones. While it is unclear how much the military has actually tested such drugs on the battlefield (for example, at the time of the Gulf War), soldiers have indulged in numerous conspiracy theories. A whole set of new anxieties is emerging as soldiers speculate on the nature of the dangerous new drugs scientists are allegedly testing on them and the paralysing side effects these have. The ultimate trauma for soldiers becomes a fight against the hidden enemy within: against those who sent him to war in the first place. ❑

***Renata Salecl*** *is senior researcher at the Institute of Criminology, University of Ljubljana, Slovenia, and Centennial Professor at the London School of Economics. Her books include* The Spoils of Freedom: Psychoanalysis and Feminism After the Fall of Socialism *(Routledge, 1994) and* (Per)versions of Love and Hate *(Verso, 1998)*

PENELOPE FARMER

# Babel in Birmingham

**'[Researching *The Snake Pit*] made me realise that it's absolutely normal to become deranged . . . that the alienation from a secure homeland [sends] so many over the edge.'**
**Olivia de Haviland**

*The speakers here all live in Oak View, a hostel for people with mental problems, in Moseley, Birmingham. As the locally notorious Palm Court Hotel, it was one of many profitable private dumping grounds for ex-patients of the asylums closed down under the policy of Community Care. Some older residents have lived there nearly 30 years.*

*With up to 70 residents, Oak View is now considered too big and institutional. Everyone, even the mentally ill, is thought better off living closer to the norm, the nuclear household. Oak View has its problems. But it works well for people too mentally frail to live alone, or not sociable enough to cope within a smaller group. Inside it, a person can both hide and have company. A fundamentally benign place, it tolerates eccentricity of all kinds.*

*Mental illness is not romantic here. There are no suicidal poets. Most residents have wretched histories, feel social rejects; some not only look but smell odd. In a few cases they make no apparent sense. But the psychiatrist RD Laing pointed out that if you really listen to the so-called mad, sense always emerges. After seven weeks living in Oak View I can confirm that. Some told their stories coherently, if obsessively. Some did not. But the stories came across anyhow. Without exception they confirmed that being mentally ill is horrible – and not as far from the rest of us as we hope. This may be the reason why such voices are rarely, if ever, heard.*

*Credit: photographs on pp49–54 by David Rhodes*

**Lilian, long-term resident** Who wants to hear about us?

**Barry (65)** I was in care from the age of three. The only time I've felt loved was when I lived with my grandma for three months when I was 15. She gave me new clothes and my brother was jealous. But my grandmother said, 'Leave him alone. You were all right, I brought you up. But he never had anything, and now it's his turn.'

**Debbie (31)** I never lived with my parents. I was always in a children's home. I used to see my dad, until one time I'd got into trouble and he took me into the bathroom and whacked me out with a leather strap. The next day I was really ill. My dad was stopped from seeing me after that. At seven I was fostered out; I was treated like a handy-cart, do this, do that. As a teenager, I was dressed up in granny clothes. All the other girls were in their tight little skirts, Doc Marten's. I hated myself.

**Peggy (70)** We lived in Birmingham. My mother wore pointed shoes and a black skirt and a white sweater. I didn't like her – she talked behind my back. She wasn't a nurse. Or maybe she was. She had to be trained to have me. It all comes down to sex in the end. Sex is bad for

women. It tears them inside. My mother had to have 13 stitches after she went with a man.

**Lilian** I don't know where I was born. I don't know who my parents were. I was in a home from nought to 16. I had my first baby when I was 19. I had nine children, didn't bring any of them up. This is an institution for people who have mental handicaps. I don't have a mental handicap. I don't want to be here.

**Alan L (42)** Do you want to know how I came here? I came down the road. (*Laughs.*) I was in homes from three years old. In the second one in North Wales the head man would have me in the office because he said I'd 'misbehaved'. He'd take my pants down, beat me on the bare buttocks with a cane then fiddle with me. It happened to a whole lot of others. When I was 15, I got involved with witchcraft. I heard a coven met up a mountain, me and another boy went and begged to join because we were so full of anger and revenge.

After I'd come to Birmingham I got involved with devil stuff again, I went to the temple in a big house near Worcester. The man who owned it, the High Priest, was nice to me. I was branded with an upside-down cross. Against Christianity, you know. They gave me some wine, and I got a bit drunk and it was put on me. It hurt. (*He shows the scar.*)

I've got bad nerves, I'm scared of the dark. I see shapes in the dark. And I see faces in my mind and mental pictures. I hear people speak in voices. But I'm a Christian now.

**Mr W (73)** Do you want to know how long I've been ill? It was January 1937 the first attack, 63 years ago. I was only 11, when life should start. But it ended for me. Four hundred of us there were in my school and I was the cleverest. I have this egotism, you see, I've always been too clever, I thought all the time of my own brain.

But there's this pressure on it. I can't do anything about it. It's physical, not just mental. It destroyed the precious friendship with my mother. What would she think if she could see me now?

**Peggy** I'm a Catholic. Protestants don't have a Virgin. Protestant girls can't be Virgins. They all wear knickers. (*She starts singing 'Onward Christian Soldiers'.*)

**Debbie** There was this man I knew from the children's home. One of they uncles. I was thrown out when I was 15, I had nowhere to go, in the end I had to live with him. He abused me. I burnt his house.

**Beryl (50)** My Dad pushed me with spiders. I hadn't done nothing. My arm was burnt. I went to the nuns after that.

**Paul (24)** I had a girlfriend for two years and my problems really started with the break. I've always been aware that I talk to men and women in different ways – that kind of splits me. I took speed and I saw a friend and his girlfriend coming towards me, and with these two people the different talkers literally split themselves, I couldn't hold them together.

Later I took an overdose. I went down to the park behind the bushes. It seemed a nice quiet place to die. But I ended in hospital, in the Queen Elizabeth. They thought I was schizophrenic, because God spoke to me once. But now they just say I've got a personality disorder. I'm not sure what they mean by it.

**Rashid (?34)** I'm a Kashmiri Muslim and a rajah. I'm not really mad, I've just been assessed as bipolar. I suffer from mood disturbance and only get aggressive when something maddens me, just like anybody else. I don't believe in Islam any more. I believe in the Jewish 'One God'. I think Jews are the best people in the world.

**Jay (25)** I first had problems when I went to India for an arranged marriage, and it didn't work out. I'd expected to get married. Just to keep family tradition going. My father's dead and I'm the oldest. But it got on top of me. And I thought, why can't I do what I want to do? The pressure is all up here. Even my family says I'm different from the rest of them. The illness first showed itself – it's the memories it brings up in me. For four years, I've suffered from manic depression. I was sectioned. I had medication. I'm off it now. Mentally they think I'm fine. It's just that . . . My mother? I argue with her.

**Peggy** My mother made me drink lots of water. She said water was good for your blood, but tea was bad for the kidneys. My stepfather was nice to me and she didn't mind. I love my mother, Dr Newman and the King of England.

**Alan H (?60)** I was a lorry driver and I was away some nights. I came home one day and there was this note from my wife: 'I've gone, don't come looking for me.' I went down the road and got a bottle of whisky – I don't know whether it was because I was sad or happy – and I drank all of it. And the next week I got a letter from the council saying I owed hundreds of pounds. I'd given her the money for the rent and she hadn't paid it. So I was evicted, I went into a residential home with my son. I was lonely. I had to choose between drinking and driving, and I chose drinking. I've been in mental hospital six times. I was in a coma for seven weeks. I died once – I was dead three minutes. I only drink Carling Black Label now. I've got a community worker, a key worker, a social worker and a nurse. I've got two doctors and two psychiatrists. I see them every month.

**Peggy** I'm drunk. I'm drunk on tea. There was the war. It wasn't a very bad war. I don't like remembering things. My mother was put in a home, I didn't put her there.

**Jason (28)** I'm hearing voices. I'm hearing my brother. He jumped under a train in 1991 and he's telling me to come and join him in heaven.

**Alan L** God and Jesus talk to me. Sometimes they tell me I'm going to hell. Sometimes they're friendly. Once I offered Jesus a cigarette. But he said, 'No, thank you, I don't smoke.' God isn't talking to me at the moment, I'm feeling too angry and rebellious. I'd like to spend all night in a graveyard.

**Jason** My mental problems started seriously when I was 18. I tried to commit suicide because I thought the end of the world was coming and that I was so evil I'd go to hell. That was when I was first sectioned and began ECT [electroconvulsive therapy] – I had 35 sessions altogether. They did it to old people too, I didn't think that was right. I wrote to the Queen about it. I was in and out of hospital, hostels, locked wards, open wards; once I was in pyjamas for three months, then my mother came in and said she'd a surprise for me – and there were my clothes.

You couldn't know what that meant if you hadn't spent three months in pyjamas.

**Paul** I've been on medication three years now. I was on Sulparide at first in the hospital. That's heavy stuff. It gave me hypermania. Some of the staff are very nice. Some of them are a bit hard-hearted. You know, sort of callous. A bit insensitive.

**Wendy (?55)** (*given much ECT*) Hello. How are you? Deaf nurses. Deaf nurses.

**Peggy** I don't like nurses. They talk behind your back.

**Mr W** I had ECT – lots of it. I had it before there were muscle relaxants, they had to hold you down to stop the twitching. It didn't help. The lobotomy didn't make any difference either. Nothing does.

**Michael (54)** In 1978, the police picked me up in Burton because I was sleeping rough. And I said I've done absolutely nothing wrong and he said, 'We know you've done nothing wrong, but we're taking you to St Matthew's psychiatric hospital, for your own safety.' So I said fair enough. And I was very, very happy at St Matthew's. I thought it was a first-class place, and in my opinion they should not have closed it down. I didn't want to leave, but after six months they had to let you go.

I'm pretty placid by nature. I've a placid temperament. I'm on tablets three times a day, for a mild case of schizophrenia, the doctor says. And apart from that I'm in good health, touch wood.

**Peggy** No more medication. No more medication.

**Reg (70, ex-tramp)** Some of the people in here (*tapping his head*), they can't speak. You have to be careful. How do you talk to them? They haven't got common sense. I shouldn't have had to come here.

**Gloria (?58)** It makes it easier to live here if you look down on people a bit.

**Beryl** (*to Dennis*) You're barmy.
**Dennis** (*to Beryl*) *You're* barmy.
**Beryl** (*to Dennis*) Of course I'm barmy; that's why I'm here.

**Eric (42)** Other people in here? They're all right; not really mad – but not quite right; just missing a screw or two. I was in hospital twice. And in prison for three months. I loved it. You're out of the world. All the pressure's off. Now? I feel aimless, haven't found my niche. I think I'm having a mid-life crisis.

**Mr W** I have to put all my effort into getting by, day to day. I can't do anything else except keep going. Having my own room. My privacy. I couldn't live with anyone, not now. I just come down and eat my meals and go away again, not talking to anyone. Doing the same things, all the time, having a routine, keeping my equilibrium, that's the one thing that helps. I would give anything for one year of just being normal.

**Dennis** We want the staff here to treat us as if we were normal.

**Robert, care worker** What's normal? Normal is vast. ❑

***Penelope Farmer*** *is a writer, now working on a book on mental health*

More articles, *Index* archive material and links ➩ www.indexoncensorship.org/madness
Put your opinion online ➩ www.indexoncensorship.org/comment
Email the author ➩ penelope.f@indexonline.org

MARTIN ROWSON

# Fools, knaves and lunatics

'In that direction,' the Cat said, waving its right paw round, 'lives a Hatter: and in that direction,' waving the other paw, 'lives a March Hare. Visit either you like: they're both mad.'

'But I don't want to go among mad people,' Alice remarked.

'Oh, you can't help that, 'said the Cat: 'we're all mad here. I'm mad. You're mad.'

You can appreciate Alice's problem. Although the narrative parameters of *Alice in Wonderland* place her in a dreamland, an oceanic maelstrom of irrationality and unreality, within the context of the hyper-reality often experienced in dreams she's constantly subconsciously conscious of being the only sane Ulysses on an insane Odyssey. But things wouldn't have been much better once she woke up. Mad hatters were proverbially mad as an occupational hazard: the mercurial steam they used to mould the material they made the hats from drove them mad as a matter of course; likewise, hares in March, in rut, behave with such abnormal abandon (for hares) that they are, by definition, mad.

But let's stick with humans. Back from the rather terrifying (if amusing) endemic madness of Wonderland, Alice found herself in a society which was also endemically mad. The free and legal availability of opiates, taken with the cocktail of chemicals the Victorians breathed courtesy of their industrial revolution, meant that most of them were, at the very least, peculiar: the cavalcade of eccentrics portrayed by Dickens are, by these lights, less likely to be irritating whimsy than rather grim documentary. And it's a small step from that eccentricity to the Victorian fondness for Nonsense. Edward Lear's fear of his epilepsy (is that a kind of madness?) led him to disguise it with a studied eccentricity that teetered on madness – he couched his only proposal of marriage in an earnest enquiry of whether his beloved could sharpen pencils: she said she couldn't, so he said 'Oh dear' and walked away – while he found comfort in Nonsense, an irrational security blanket to clutch in the face of an unforgivingly Rational world.

After a career depicting the madness of the world, the great satirical caricaturist James Gillray is believed to have leaped to his death from his garret window above Mrs Humphrey's print shop, a fortnight before the Battle of Waterloo and eight years after he'd sunk into madness himself. He was the only one of six children to reach adulthood, and was brought up as a Moravian, a Protestant sect that viewed the world with horror and welcomed death as a (literally) blessed release; as he got older he suffered increasingly from morbid depression and was growingly obsessed by his failing eyesight, a condition exacerbated by his prolonged exposure to nitric acid, a chemical central to the etching process. Did that help him, like the Mad Hatter, go mad, or was it the circumstances of his childhood? Whatever the cause (and we shouldn't forget another kind of occupational hazard, the savage intensity with which he chronicled a Mad World driving him, like Swift and Goya, mad too), during a period of brief lucidity in 1813 Gillray gave an audience to Mrs Humphrey's latest protégé George Cruikshank. The purpose of the audience seems to have been a sort of satirical blessing, a kind of caricaturist's apostolic succession, but all Gillray would say was: 'You are not Cruikshank, but Addison; my name is not Gillray, but Rubens.' Cruikshank went away unblessed, but later acquired Gillray's table if not his madness. Later in his career, when most people dismissed him as irredeemably eccentric, Cruikshank became a warrior for Temperance, the Victorians' very own War on Drugs, against a self-imposed and not always temporary madness. Indeed, they called it 'Drink-madness', a blight on both productivity and decent morality, even though, in his will, Cruikshank left his wine cellar to his mistress.

No human society has ever existed without some psychotropic or mood-altering mechanism to allow us to look at the world in a different light from the harsh and unbearable glare of Reality, be it booze, fags, dope, chocolate, Dionysiac frenzies, political monomania or just sitting still and meditating on the unknowable infinite. I'm told that if you don't eat for a fortnight you get wonderful visions, as religious mystics have for centuries. You can achieve the same effect with Benalyn expectorant and vodka chasers. Gillray's contemporary William Blake seems to have achieved this without the outside catalyst, and is now universally recognised as the greatest English visionary, offering us sight, 200 years later, of a different, mystical, spiritual England in opposition to the tyranny of Reason we're currently in thrall to. But look at his work, at those tiny, tiny printed pages (produced and coloured in the same poisonous miasma as Gillray's) crammed with text which then curls, madly, up into the margins to hammer the elusive point home. This is the textbook stuff of schizophrenia.

But so what? About 20 years ago, an article in the *British Medical Journal* deplored advances in the treatment of syphilis because the extirpation of general paralysis of the insane, a frequent symptom of tertiary syphilis, denied our antiseptic world the mad genius of Van Gogh, Nietzsche, Schubert and many others. Unkissed by Venus, the Victorian painter Richard Dadd did his best work in Bedlam, after going mad and killing his father with an axe. The cat painter Louis Wain ended up in the same place, explaining only with difficulty to a passing visitor that not only did he paint like Louis Wain, but he *was* Louis Wain. 'Of course you are,' purred his well-meaning interlocutor. Earlier, of course, Hogarth's Rake ended up in a different Bedlam elsewhere, as an awful warning to the rest of us, before Madness came to rank equal with Death as an exquisite and slightly delicious Romantic fate. Think of Ruskin and the first Mrs Rochester. Much later, think of hippie Romance and the rock 'n' roll martyr Syd Barrett, the founding genius of Pink Floyd, still alive but lost to us for ever after frying his brain with LSD, opening the Doors of Perception and thereafter drawing a blank.

And let's finish with mad Dean Swift writing, in *A Tale of a Tub*, his 'Digression concerning the Original, the Use, and Improvement of Madness in a Commonwealth', where madness, as manifested in political megalomania and bellicosity, is equated with an excess of semen being diverted to and infecting the brain or the inability to have a damn good shit. The gag – the ironic point – is that the inmates of Bedlam would function perfectly well in the law, medicine, the Church and politics if released into the outside world. In between the ironies, however, is Swift's true lesson, which is tolerance: it's the truly mad who, through philosophy, religion or politics, seek to make everyone the same as them. In the face of this universally prevalent Madness, Swift advises that we seek 'the serene peaceful state, of being a fool among knaves'. ❑

***Martin Rowson*** *is a cartoonist and satirist* *(see p108)*

To H. Fuzelli Esqr this attempt in the Caricatura-Sublime, is respectfully dedicated.
WIERD-SISTERS; MINISTERS of DARKNESS; MINIONS of the MOON.
"They should be Women! — and yet their beards forbid us to interpret, — that they are so."

*Page 58, above:*
*William Hogarth,* Scene in Bedlam, *the last episode of* The Rake's Progress, *engraving, 1735. Credit: Wellcome Library, London*

*Page 58, below:*
*James Gillray,* Wierd-Sisters, *etching, 1791: a parody of Fuseli commenting on the madness of King George III in 1788–79*

*Page 59:*
*William Blake, illustration for* The Book of Urizen, *etching, 1794. Credit: AKG Berlin*

*Left: Richard Dadd (1819–87),* Portrait of a Mad Lady. *Credit: The Bridgeman Art Library*

*Above: unknown artist, engraving of the 'mad' Marquis de Sade (1740–1814) in prison. Credit: The Bridgeman Art Library*

*Right: John Tenniel,* The Mad Hatter, *from* Alice's Adventures in Wonderland *by Lewis Carroll, engraving, 1865. Credit: The Bridgeman Art Library*

*Right: engravings by Louis Wain (1860–1939)*

FORTUNE-TELLING.
"You will get a great surprise."

" How do you account for your wicked temper, Girlie !"
"You ! Boy mine ! "

*Left: Dr Hugh Welch Diamond,* Depictions of the Insane, *1856. Credit: Royal Society of Medicine, London*

ADEWALE MAJA-PEARCE

# Beaten, manacled and raped

**Contrary to the assumption that attitudes to madness are more benign in traditional cultures, the lot of the insane in Nigeria, at least, is not a happy one**

Over ten million Nigerians – 10% of the population – are reckoned to be suffering from one form or other of mental illness. A small minority of them receive orthodox or 'western' medical attention in one of the eight Federal Government-owned neuro-psychiatric hospitals in the country, but a more significant number are treated by traditional or native healers, especially in the rural areas where poverty, ignorance and the absence of even rudimentary healthcare clinics are commonplace.

Many of these native healers are freelance operators; others belong to the Nigerian Association of Medical Herbalists. These native healers have a widespread reputation for excessive brutality in the treatment of especially difficult patients. Henry Williams, a US psychiatrist who undertook research in the country in the mid-1980s, spoke with many patients who described 'chaining and flogging' as common. He himself observed 'many patients whose wrists and backs showed scarring and ulceration from alleged manacling and beatings'. On a number of occasions, I have myself seen manacled patients being led through the streets by these native healers, but I had never before entered any of their establishments. But neither of the two I visited in Abeokuta, home to West Africa's most famous neuro-psychiatric hospital, were the dens of horror I had imagined.

The first was a purpose-built 'hospital' – Ademola Mental Hospital – which consisted of two single-storey buildings with six rooms each facing each other in a clearing in the bush on the outskirts of the town. The native healer was a slight, softly spoken man in his early 40s who readily admitted that he sometimes flogged his patients, although he reserved that drastic measure for the ones who were already far gone: 'the stubborn ones in the market place' was how he put it. Flogging them was the only way to get their attention. He didn't have any such at present, he said, indicating the three patients on the opposite verandah, one of whom was manacled by his ankle to a wooden post. The man himself didn't seem perturbed by his condition or cowed in any way and never stopped smiling at me the entire time I was there, in contrast with the other, older man who kept pacing up and down and muttering to himself while a nursing mother sat quietly on a raffia mat suckling her child, seemingly oblivious of the activity around her. The native healer said that the man had been manacled at his own request not two hours ago because he thought he was about to have one of his turns. The last time he had one of them he got as far as Kaduna, nearly 1,000km away, before he came back to himself. He would remain manacled for another two hours, by which time the concoction of herbs he had given him would have taken effect.

The other native healer I spoke with, an Alhaji in his late 60s who lived in the town proper and worked out of his home, denied that he ever resorted to beating and, to prove it, left me alone with the two patients he was currently treating while he busied himself at his desk at the far end of the room. I spoke with the man first. He said that he was 45 and a house painter and that he started exhibiting signs of madness in his late 20s. He said that whenever he was about to fall sick it was like ants were scurrying across his brain and then he would blank out. His family would later find him wandering the streets, although he never went naked like some of the others, he quickly added. He said that his current treatment, which consisted purely of herbs, had lasted five years but that he was now fully cured. He was looking forward to going home.

I turned to the woman, who had all the while been sitting quietly beside him, her hands nestling on her lap, her legs tucked under the hardback chair and an air of sadness about her pretty, delicate features. She said she was 28 and a nursery school teacher and had first started exhibiting signs of madness ten years previously. 'I have brain disorder,' was how she put it. She said that she would suddenly 'be feeling headache' and then she would start talking to herself. She said that she had been with the Alhaji for six months and had been cured in the last fortnight. She was anxious to resume her career, she said, and live a normal life.

When I was done, the Alhaji showed me his laminated membership certificates. Like the other native healer, he said he had inherited the business from his father and he used only herbs in his treatment, some for drinking, others for bathing. He said he didn't believe in tablets and injections, which offered only a temporary solution, which was why some psychiatrists at the orthodox hospitals often referred patients to him. I asked him whether he could treat all mental illnesses and he said yes; you only had to know the right combination of herbs. I asked him how many types of mental illness there were and he explained that some were hereditary, some were caused by excessive smoking of marijuana and yet others by a particular type of wind that you met walking down the street. But he said that about 80% of all insanity was caused by people with evil intentions. These witches had the power to make you mad by ejecting an invisible powder from the tip of their tongue when you turned in their direction, especially if that particular wind happened to be blowing.

The belief in witchcraft as the root cause of insanity is strong in Nigerian society and is related to another superstition that poses great risks to the lives of women lunatics: that impregnating such women is a sure route to riches. One man, who 'collected' lunatics 'as a humanitarian job which I have to render . . . my fellow human beings', and who confessed to chaining them by their wrists and ankles for up to a fortnight at a time whenever they got violent, complained of the incessant demands from 'big, big men' wanting to borrow them for the night. Only the previous day, he said, 'a complete gentleman' offered him 'big money' to take 'one of my mad women' to a hotel, although he always turned them down. He added that he himself refrained from sleeping with any of them because 'it is sinful to do that', but that whenever he felt the urge he would go to a nearby hotel, 'cool myself down with two bottles of stout, a stick of cigarette and make love to one of the prostitutes'.

Easier still is to pick one of the insane woman from the streets. The *Nigerian Tribune* reported a case in 1998 involving a councillor in Owo in Ondo State who 'prostrated before the elders for forgiveness' after confessing to having impregnated a well-known lunatic in the town known simply as Tola. His excuse was that the Devil pushed him into it when he was drunk, but eyewitnesses claimed to have seen his car collecting her on more than half a dozen occasions some months previously. 'Pregnant lunatics roam streets' was a recent headline in the *National Concord* following the sudden appearance of 'female lunatics aged between 18 and 25 [who] roam the streets in search of daily bread'. The State Commissioner for Woman Affairs and Social Development Programme, the government department responsible for caring for these woman, opined that 'it takes two to tango', and added that 'if the men . . . in their desperate search for money had not slept with the female lunatics they wouldn't have been impregnated.' Not that they need all have been the victims of greed: in the anonymity of Lagos, female lunatics are casually gang-raped by the local 'area boys' because, according to one of them, 'they are very easy to get and it is free of charge'.

The desperate plight of the insane in Nigeria is borne out by the existence of only one of the 100 or more NGOs dedicated to their welfare, the Mental Health Rehabilitation Foundation (MHRF), and by official indifference, as recounted by an MHRF spokesperson. 'They gave us N10,000 (US$100) five years ago. That was the last we heard of them. We visit their offices and they keep on promising. As with Nigeria of today, nobody can challenge them for not doing their job and, when we persist, they tell us to go and look for other jobs rather then lending a hand to charity. They even say they are not the ones who asked these people to go mad.'

This official indifference and even hostility extends to the treatment of insane prisoners. In an unpublished manuscript, Kunle Ajibade, one of five journalists incarcerated for coup-plotting by the late dictator, General Sani Abacha, whose own brand of lunacy threatened to plunge the country into another civil war, recounts the pathetic story of an insane prisoner who was flogged in order to establish whether he really was insane, and then flogged again for being insane.

> 'I am digging for gold. Digging. Digging for gold.' The noise was so deafening that midnight. It woke up many of the inmates. The noise-maker: a madman, who was brought to Makurdi prison merely for stealing some clothes belonging to his brother's wife. He was locked in the punishment cell. With all the shackles on, he dug the floor of his cell with his fingers.

'Nobody should join me-o. If you join me I will break your legs.' And the warders on night duty laughed. 'Matthias, Matthias, my boy. Bring my food. With plenty meat. I don't want water-o. Only wine.' And that in the dead of night. Again, the warders laughed and laughed.

The following morning, he was dragged out of his cell on the orders of the man in charge.

Thinking that the madman was pretending, the man in charge ordered that he should be flogged. He then asked the madman, 'Do you know me?'

'Yes, I know you,' the madman replied.

'Where do you know me?'

'I know you, now.'

'I say, where do you know me?'

It took the madman a moment before he said. 'You are a tout. I know you in the motor park. No be so?'

The officers dispersed. They had confirmed that truly Abraham was mad. For many nights and days they tied him to one of the trees in the yard like a dog. I was traumatised by the frequent beatings of that madman. ❑

***Adewale Maja-Pearce*** *is a freelance writer who can be reached at majapearce@hotmail.com*

**SARA MAITLAND**

# The changing shape of madness

**Because they are ill, and because society finds them threatening, the mad are deprived of full humanity. One way in which this is achieved is the denial of their right to be heard. By refusing to listen, we silence them**

If a tree falls down deep in the forest where there is no one to hear it, does it make a noise?

This well-known Zen *koan* depends on the realisation that there are two sorts of silence. There is the silence when no one utters and there is the silence when no one hears. When a radio receiver is turned off, the radio waves still exist, but the effect is of a silence identical to the silence there would be if the radio stations stopped transmitting.

Of course it is possible to stop someone speaking. The most efficient way of doing so is to kill them. Or you can frighten someone into 'voluntary' silence; render it physically impossible for someone to speak by, for instance, cutting out their tongue or cauterising their brain; impose such powerful cultural taboos that something quite literally becomes unspeakable – as, for instance, the name of God in Jewish religion.

All these methods have their problems. A more efficient method is to damage the receptors; to prevent the utterance being audible. Much censorship works this way – by preventing the would-be speakers from using any channel of public communication. If you put people in solitary confinement, for example, you have not prevented them from speaking; they can, and by all accounts often do, shout a lot and make a good deal of noise: what you have done is prevent them from being heard. You have silenced them.

There is, however, an alarming tendency in the silenced to develop new and creative channels of communication. Procne had her tongue cut out to prevent her revealing her brother-in-law's incestuous and adulterous rape, so she took to embroidery; Russian dissidents generated *samizdat* texts; prisoners in isolation cells tap on metal pipes.

There is a more subtle way of achieving this type of silencing – you can set up so much 'static' or interference around the speakers' discourses that although other people can physically hear the noise, they cannot make any sense or meaning of it. One of the most potent forms of static is to diagnose someone as suffering from a mental illness.

'Mental illness' is a complex, and highly contested, phenomenon. It appears to be fully described with elaborate tabulations of discrete diseases, and distinctions. Careful differences are drawn, for instance, between psychosis and neurosis, between schizoid and affective disorders, between treatable and non-treatable conditions. But in fact there is no profound agreement about any of these categories, nor about their causes or their trajectories. Nor, despite a plethora of treatments – pharmaceutical, psychological, mechanical, behavioural – is western medicine at all competent at curing them.

This is not for lack of effort. The debate about the nature of mental illness began very early in western history, in the classical period, with a profound, and frequently abusive, disagreement between three interpretations: a 'medical' model, advocated by Hippocrates; a philosophical model, championed by, among others, Plato; and a moral/pragmatic model put forward by the Stoics. Two and a half millennia later, we still have these three schools of thought – 'keep taking the tablets,' 'talking cures' and 'pull yourself together.'

On the whole, though not, alas, entirely, we have abandoned a fourth causal model, highly popular during long swathes of western history: the demonic possession model. This had both a 'moralistic' version (the Devil takes his opportunities) and a 'bad luck' version. The decline in popularity of this narrative is not entirely due to increased rationality and a decline of religious faith. In popular culture, at least, we have hived this model off from mental illness, and apparently believe we can easily and clearly 'distinguish' between the mad and the bad; the ill and the evil.

What the various schools of thought did not contest is that mental illness is real illness, causing, as do other illnesses, real pain, real grief and real affliction. Despite some interesting interventions during the second half of the last century from what is commonly called the anti-psychiatry movement, it is cause and treatment that are contested, not the event.

The 'static' arises because mental illness is not like other illnesses. Although we use similar medical taxonomies, 'schizophrenia' is not like 'measles' or 'appendicitis'. It is not even like AIDS or ME or other similar 'syndrome' diseases. Mental health diagnoses are made, necessarily subjectively, from lists of 'symptoms' which have no external verification and may depend heavily on the questions asked. If you asked an evangelical fundamentalist, a liberal Roman Catholic, a conservative agnostic and a Buddhist if they felt 'deserving of love', you would need to know a good deal about *theology* to diagnose whether they were suffering from low self-esteem, delusions of grandeur, paranoia or healthy cultural/social integration.

So the diagnostic tools for mental illness become a list of behaviours and opinions and, obviously, these are highly socially determined. Refusing to take off any clothes at all on a private beach in the Caribbean and insisting on taking them all off on an inner-city street could both be offered as diagnostic symptoms (repression/exhibitionism).

Moreover, the 'symptoms' on the diagnostic lists change – behaviours being added and subtracted according to social (rather than medical research) criteria. Most famously, 'homosexuality' was a classified symptom of mental illness until, one day in 1973, the American Psychiatric Association voted by 58% to 38% (4% abstention) that it wasn't. Did all those 'ill' people get better suddenly – abracadabra – or had they never been 'ill' in the first place? Usually, the changes are more gradual but, therefore, in a sense more subjective: 'appropriate gender adjustment', for example, has undergone continual, gradual change in my lifetime. In the 1950s, women could still be institutionalised for having illegitimate babies; now women who remain virgins until they marry are considered somewhat 'odd'. Currently, the status of Voice Hearing as a clear schizoid symptom is under intense scrutiny in some psychiatric circles while in others it remains a stable and authoritative diagnostic determinant.

There is, inevitably, a socio-political dimension to this too. In the 1950s and 60s, a very large number of women were prescribed Valium as a tranquilliser. This turned out to be damaging in various ways and highly addictive, and the patients started to complain. Prescription rates dropped markedly – and a new 'illness' emerged: ADS (attention deficiency syndrome). Now we have a swiftly increasing number of children being prescribed Ritalin as a tranquilliser. It has some marked advantages – it is a great deal cheaper than better housing and smaller class sizes, and the 'patients' are not in a position to protest or complain.

Black, especially Afro-Caribbean and especially young male individuals, are more likely to be sectioned, more likely to be diagnosed as psychotic, more likely to be more heavily medicated and less likely to be offered 'talking' rather than prescription therapies than the white population of matched economic background. In a large-scale international study, the World Health Organisation concluded there was no geographical or ethnographic difference in schizophrenia rates. Given the cultural determinants of the diagnosis this is massively problematic, but nonetheless suggestive.

The point is that we have a category of 'illness', undeniably a painful and therefore authentic illness, within which it is possible to construct, insert and extract a great number of different kinds of behaviours as symptoms: including the denial that one is ill. There is a socio-political dimension to all health issues; but mental illness offers a special and particular mechanism for social control. It is established as an illness, requiring both management and compassion, but its diagnostic criteria remain extremely flexible, socially constructed and amenable to manipulation – conscious and unconscious.

This is a 'medical' narrative of madness as illness. But there are other ways of decoding insanity. Alongside this compassionate 'scientific' model there are other cultural accounts: for example, of madness as the 'price of genius':

> Great wits are sure to madness near allied
> A thin partition do their bounds divide.

Or of the mad as touched by the gods; cursed by the gods; subhuman and 'animal' and/or 'more in touch with nature'.

These sorts of cultural constructs have as long and complex a history as the medical models do. Longer, perhaps, because they are embedded so deeply in myth, religious ritual and preliterate social custom. They, too, have roots deeply dug into the classical world. We may choose to see Hellenic culture as the model of the civilised and the rational; but

Herodotus reports that the Scythians, contemporaries and neighbours of the Greeks, blamed them for all the insanity in the world because it was not an act of reason to adopt and worship a god who drives humans mad. They were referring to the worship of Dionysius, whose rituals did indeed drive his devotees to hysterical, ecstatic or manic states, and whose cult was extremely popular. But there was a wider appreciation of the 'double nature' of lunacy: at Delphi, the Pythia, the deranged, ranting, incomprehensible prophetess, established direct contact with the gods – and the Greek states were prepared to pay very highly for access to and interpretations of her oracles. Orphic mysticism can be hard to distinguish from florid psychotic utterance. In short, the mad were, in many instances, as sacred in rational Greek culture as in any other pre-scientific society. The early and medieval Church took up this theme, elaborating ways to tell whether someone was ill, demonically possessed or a direct vehicle of the divine. At times this must have made life pretty perilous for the less stable: it was deemed heretical and therefore a capital offence to be mad, but apparently identical behaviour could get you canonised – in the case of Joan of Arc it did both.

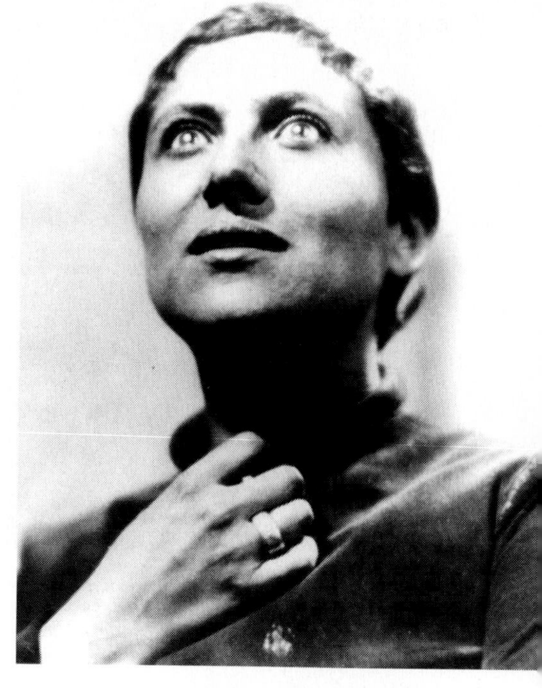

*Maria Falconetti as Joan of Arc,* La Passion de Jeanne d'Arc, *dir. Carl-Theodor Dreyer, 1928. Credit: AKG London*

The Renaissance, asserting the creative and 'non-conventional' authority of the individual and especially the artist, privatised this strand of thinking with the idea that madness was the price of true sensitivity and therefore related to genius. Hamlet becomes a hero instead of the butt of endless Chaucerian-period jokes about the ineffectual mummy's boy who can't make up his own mind. The idea of the creative 'muse', who could descend upon the individual who was too profound to be satisfied with life, was somewhat similar to ideas of Dionysian possession

*The Ecstasy of St Theresa*, sculpture by Giovanni Bernini (1598–1680), Santa Maria della Vittoria, Rome. Credit: The Bridgeman Art Library

– and without this possession there would be no true creativity. Even a highly disciplined puritan like Milton, whose work ethic was so strong that he provided Adam and Eve in Paradise with light gardening chores since true idleness would have been less idyllic than some useful tasks, could write of melancholy as a 'Goddess, sage and holy'.

The Enlightenment endeavoured to banish the irrational, the superstitious and the mystical. The very real advances in scientific understanding and outlook required even the word 'reason' itself to change its meaning. Seventeenth-century divines could write that reason was 'a box of quicksilver . . . a dove's neck or a changeable taffeta . . . a floating island', or 'a candle of the Lord' – an instrument to decipher the revelation of God in the created order. The Enlightenment rewrote that interpretation so that reason became not constructed but self-generated, a tool for calculation, classification and systematisation. Above all, 'reasonableness', rationality, became the mark that distinguishes humans from other animals (replacing the previous religious paradigm: that our primary humanity was separated from animal nature by Christ's assumption of humanity in the Incarnation). This meant that, although illness was much better understood, mad people were treated appallingly, as they had, by the nature of their condition, become less than fully human – along with women and non-Europeans; though both these were, at least, amenable to training. Early in this period, Spinoza declared that 'everyone has the power of clearly and distinctly understanding himself and his emotions, and of bringing it about that he should be less subject to them'. If you were mad, if you could not be 'less subject to your emotions', you were either a self-indulgent disgrace or less than fully human. The infamous leisure activity of watching the Bedlam lunatics was but an equivalent of taking the children to the zoo.

Romanticism endeavoured to reinscribe some types of mental illness – depression, terror, mania, addiction, and some forms of depression in particular – within the category of extreme feeling called the 'sublime'. The mad became sacred again, a sort of superhuman, glamorised and idealised. This, coupled with nineteenth-century philanthropic ideals and a growing belief that some lunatics could be cured, gradually led to an amelioration of the treatment of the mentally ill, but it did not seriously challenge the status of madness as 'other' – as outside the boundaries of the socially human.

If my first (highly simplistic) survey was meant to suggest the ways in which mental distress has been constructed as illness, then this second skim through three millennia is intended to hint at the ways in which madness has been constructed as liminal space. The mad live, or are perceived as living, on the boundaries: on the boundary between the wild and the civilised; between the human and the non-human; between the rational and the divine. Liminality is, of its nature, profoundly disturbing. Hybrids, wild children and 'monsters' all both fascinate and repel: the enduring fairy stories of changelings, mermaids, vampires and werewolves; the myth of Frankenstein's perilous experiments; the history of the freak show; the current social anxiety about cloning and robotics; the terror of aliens; the popularity of science fiction – all attest to the rabbit-and-the-snake enchantment that liminality casts. Madness provokes very strong feelings of unease. Our present society is, despite all the talk of 'political correctness', highly proficient at excluding that which makes us uneasy. The old, the physically handicapped, the foreign, the Other – anything that might challenge the notion that we are good and beautiful and will live for ever and have a right to all those things.

Stigmatising is very easy. The insane present a particular danger, however, because you can't always tell who is mad just by looking at them: it could be anyone, so could it be me? They are there, in our midst, and they are frightening. The most common reaction to manifest madness is a mixture of revulsion and anxiety. It therefore feels necessary to draw a clear line between the 'mad' and 'me'. I may have my little ways, but I am not mad. The boundary must be marked, must be stabilised, must be made clear. Madness is a threat to good order and to my sense of myself as human and rational. Madness is dangerous.

This sense of outrage and fear is given some justification because sometimes some people who are mentally ill, as well as doing unconventional and disturbing things, also do some dangerous things. We need to keep a sense of proportion about this. More fires are started by mismanaged chip pans, insurance fraudsters and naughty children than by pyromaniacs. More people are killed by drunken drivers than by schizophrenics. More children are hurt by accidents in the home than by aggressive paranoiacs on the street. More children are abused and more women raped by members of their households than by sex maniacs driven by uncontrollable Voices. As a matter of measurable fact, in contradiction to the impression given by our tabloid media, the number of murders committed annually by individuals with a mental health diagnosis is not increasing; it has actually declined as a percentage – because more 'sane' people commit murders. None of these facts is controversial; but they are not sufficient, nor sufficiently proclaimed, to overwhelm our fear of the liminal, the Other, the uncertainty about our own boundaries.

As well as doing dangerous things to other people, there is a serious concern about 'self-harm'. A tragic number of people with a diverse range of mental illnesses do commit suicide, as well as inflicting serious though less final damage on themselves. But we are in a strange cultural place when we argue for euthanasia and even assisted death as a 'right' and yet believe that someone can and should be deprived of basic civil liberties – detained and forcibly medicated – because they are a 'danger to themselves'.

This is not a new problem. In classical mythology, Cassandra, a princess of Troy, was given the gift of foretelling the future; she was also given the anti-gift of no one ever believing her. For the ten long years of the siege she went about announcing that Troy would fall and everyone said she was mad. If to be mad is to be non-rational, and it is rational to tell the truth, then obviously the mad can't tell the truth. But lots of rational people don't tell the truth. And lots of madness, even severe psychosis, is 'partial'. That is to say, I might be mistaken in believing I am made of glass, or that Prince Philip was trying to seduce me, but be entirely correct in believing that I am being bullied by the staff on my closed ward. Yet the two become elided.

This should have been the significance of Dr Schreber's case. Daniel Paul Schreber, an eminent German judge, first went insane in 1884 and was confined to an asylum. He was placed under 'tutelage' (a legal order to protect individuals from negative effects of their own actions, including unwise financial administration) in 1894; this was confirmed in 1900, but Schreber appealed. He had a highly florid mental life that most people would find fairly peculiar, not to say delusional. His mood swings were extreme and his behaviour was anti-social and genuinely distressing to others. Let us for a moment accept the diagnosis of florid psychosis: he was 'acutely mentally ill'.

The administrator, and senior consultant, of Schreber's asylum, Dr Weber, claimed that Schreber was 'influenced by hallucinations and delusions', 'no longer master of his own free will' and therefore not fit to cope with his own business affairs. (There is no evidence that he wanted to defraud Schreber, but equally there was no therapeutic motive in his deposition: for him it was ideological.) Schreber energetically defended his own interests, arguing, eventually in court, not that he was sane, but that the fact that he was mad, and held unusual religious beliefs, did not justify the court in depriving him of his financial independence and his legal rights. He won his case. There was no disastrous outcome. Even if there had been, he would have argued that he had the same right as anyone else to mismanage his own property.

As part of his defence, Schreber wrote a *Memoir* describing his own condition and his beliefs. Unfortunately, the excitement caused among the nascent psychoanalytic community by the bizarre contents of this work rather obscured the importance of his legal victory: severe insanity did not necessarily render a person unfit for citizenship, let alone other human rights. The combination of education (enabling him to produce the *Memoir*) and financial resources (giving him a recognisable interest) meant that he was in a position to contest his own silencing. In most of Europe today, however, no court would hear such a case.

I am trying to argue a double silencing here. The symptomology of mental illness together with the cultural constructs we have placed upon it – especially the prevailing view that 'rationality' is key to full humanity – leave the sufferers of these illnesses particularly vulnerable. They are a danger to themselves and others, and society therefore has both a right and a duty to control them, to force them to 'get better' and to receive treatment without consent. Part of this treatment will inevitably be to reduce their desire and capacity to resist treatment.

There is a crafty circularity in this argument: refusing consent and resisting treatment, along with 'non-compliant behaviour', become proof of the illness and a justification for acting without consent and for compelling compliance.

It is important to remember that 'danger' is implicit – like the old UK 'sus' law under which a person could be apprehended by the police merely on the 'suspicion' that he might be up to no good, and which was withdrawn because of its abuse within a racist culture. Imagine the outrage among drivers if the police were allowed to charge someone for 'looking as though they might be going to drive over the speed limit'.

Modern drug treatments can now be administered effectively without committing a person to a residential hospital – this is why the proposed new Mental Health Act will construct non-compliance with a drug regime 'in the community' as an offence. This extends control into new areas (at present compulsory treatment requires hospitalising), while withdrawing protection from an ill person obliged to live in a society notoriously unsympathetic to the illness and its sufferers.

The same symptomology and construction also leaves individuals with psychiatric diagnoses extremely vulnerable to both legal and personal abuse, well beyond the purlieus of the illness itself. It is understood to be the nature of their disease that they are irrational and therefore unreliable. They are lacking both 'in sight' – a semi-technical term, often used as a diagnostic symptom, meaning an understanding of their condition and thence their identity and situation, which is better understood by their care providers – and in what might be called 'out sight' their perceptions of the world around them are deemed to be faulty. Nothing they say need be heard too seriously.

People in severe depression, people bombarded with auditory hallucinations, people wrestling with the problems of intense physical delusion, for example, and people heavily medicated as a treatment for these conditions have, perhaps not surprisingly, the greatest difficulty in speaking energetically in their own right. When this speech is deemed 'inaudible' because it is inappropriate, frustration and anger may very legitimately mount. Attempts to speak become less and less 'logical' and may well become aggressive, florid, excitable or inarticulate. To speak 'logically' becomes a requirement if you wish to speak at all. Interestingly, to refuse to speak (voluntary mutism) *and* to speak outside the fairly limited framework of 'logical' syntaxes are both regarded as symptoms of mental illness. There is no real reason to assume a lack of coherent meaning, rather than the hearers' incomprehension, in even the most florid psychotic utterance; what it does is place the burden of 'comprehensibility' on the person the hearers have decided is least able to exercise it. At this level, the refusal, for instance, in classical psychoanalysis to 'treat' psychotics – as opposed to neurotics – is somewhat suspect.

Within this framework, the failure to use logic 'properly', or to speak appropriately, is then deemed to be yet another symptom that renders whatever they wish to say even less audible. The mad are silenced.

The second half of the danger is better known and has been systematically and bravely exposed by defenders of free speech. If what the mad say is nonsense and does not need to be heard, then anything that we would like to be nonsense and do not want to hear can, too easily, be named mad. The Lucan gospel writer has Jesus say almost jokingly, 'John came to you fasting and you said he was mad; I come to you not fasting and you say I'm a drunk.' Two quite specific categories of speakers whose words can still be treated as inaudible.

Because of the nature of mental illness and the impossibility of objective 'scientifically' measurable diagnosis unbiased by cultural considerations, compulsory mental health orders will remain an effective instrument of oppression, censorship and silencing for any society or government that wishes so to use them. It is more effective than imprisonment, because in sophisticated hands you can actually use the 'treatment' to 'change someone's mind'. Really efficient silencing has the 'speaker' agree that what he was saying was mad, and he is now better. It is also more effective because people are more aware of the way the law can be manipulated than they are of the way psychiatry can be (this is probably because lawyers are less easy to silence than mental patients). I do not feel it necessary to run again through the sickening abuses of psychiatry in the Soviet Union and elsewhere during the last century (pp81, 92). There is widespread recognition of a history of using both mental health diagnosis (he is mad, so we can lock him up) and psychiatric treatment (she is mad, so we are curing her with mind-altering drugs/electric shock treatment/etc.) as a form of censorship and silencing.

I do not want to suggest that either group – the medically mad or the politically mad – have a priority of need. Nor do I want to suggest that there is an obvious or simple solution: no developed society has yet found a way of dealing with the social problems of mental illness without some element of compulsion and detention. But it is clear to me that the social construction of mental illness, and the consequent stigmatisation and silencing of the sufferers of these illnesses, is precisely what makes psychiatry available as an instrument of direct political censorship and silencing. ❑

***Sara Maitland** is a novelist and commentator with a regular review column in* Open Mind. *Her radio play* Other Voices, *on the experience and history of aural hallucinations, will be broadcast on BBC Radio 4 in October. She is working on a cultural history of silence*

## Cleaning up protest

The elaborate preparations made by the city of Genoa in the hope of containing protest at the G8 summit in July this year are well known, as is the brutal treatment meted out by Genoese police and security services to anti-globalisation demonstrators.

Rather less familiar is the use of psychiatric hospitals in the city's plan for containing any disorder. The Genoese authorities sent a special order to all the hospitals of the metropolitan area, to be applied during the days of the G8 meeting. It ordered the use of 'forced sanitary treatment' for any protester who protested at their treatment by the forces of law and order.

Forced sanitary treatment is a form of psychiatric treatment prescribed only in the most grave cases and can be imposed only by order of the mayor with medical consent.

An eyewitness provides the following account of one such case:

> On that Sunday, in a beaten and humiliated Genoa, I was at the San Martino hospital looking for a friend who had felt on his own body the violent arm of the law. Roberto Sarlo is a videomaker and people's artist whose bruised and swollen face was displayed in *Manifesto* on Sunday 22 July. He was brutally beaten by five policemen and was treated at the first-aid station. On his release, he was immediately taken to the psychiatric ward of the hospital and subjected to forced sanitary treatment. I found him stuffed full of drugs – forcibly administered against his will. He was knocked out by a super-potent mix of Serenase and other junk.

***JVH***

**DANNY SCHECHTER**

# Is Falun Gong going crazy?

**China gives a new lease of life to the old Soviet practice of silencing dissidents by certifying them**

Throughout the history of protest and resistance movements, people in power have denigrated their opponents with hostile language and repressive reactions that demonised their image, damaged their credibility and misrepresented their motives. In the USSR, they first declared their dissidents 'insane' and then locked them up in asylums to silence their voices.

China is currently reinstating the practice. Falun Gong practitioners are being castigated as crazy and tossed into mental hospitals. Borrowing the old Soviet practice, Beijing is upping the ante with a far higher number of people falsely diagnosed as mentally ill. There were protests worldwide when the USSR resorted to this attack on its dissidents, largely because prominent writers and well-known critics were involved, and some were released (p92).

The Falun Gong practitioners are less well known and any intervention on their behalf is conditioned by the West's policy towards China; this is driven by economic issues, not ideological divisions. As a result, governments have said little about China's treatment of the Falun Gong out of a desire not to antagonise China at a time when its economy is growing while that of the US and Europe is contracting. US secretary of state Colin Powell did not raise the issue during his 29 July visit to Beijing.

One reason for the lack of protest is undoubtedly the lack of media attention. While Falun Gong protests get more coverage now than earlier, there is a lack of investigative reporting by Western media organisations. One journalist who did such work, Ian Johnson of the

*Macau, China, 1999: police carry off a Falun Gong member during a protest.*
*Credit: Mike Tam / Camera Press*

*Wall Street Journal,* won the prestigious Pulitzer Prize for his efforts – but he was quickly transferred out of Beijing. Few if any of his colleagues took up his muckraking interest in the story.

In the US, besides the practitioners themselves, there is only one lonely but credible voice being raised in protest. Dr Abraham Halpern, a professor emeritus at New York Medical College and one-time civil rights worker who worked with Martin Luther King in Alabama in 1965, has taken the lead. He told me he believes 'the [Chinese] government needs to hospitalise wrongfully dissidents who are not mentally ill because this will help them in their effort to paint the Falun Gong practitioners as not being against government policy but as mentally ill. Even if they were to hospitalise only a small number, word would soon spread that Falun Gong practitioners were crazy.

'Deliberate hospitalisation, wrongful hospitalisation, is only part of the problem. They then make it very difficult for the practitioners to get out of the hospital by demanding that their families pay exorbitant amounts of money for their "treatment" in the hospital. There's no question that this government-sanctioned conduct is a serious violation of human rights. And we'd like to stop it before large numbers of dissidents are incarcerated in hospitals as they were in the Soviet Union.'

Halpern is lobbying professional organisations. The Committee on Misuse and Abuse in Psychiatry passed a resolution asking the American Psychiatric Association leadership to ask the World Psychiatric Association to investigate this problem as it did in the USSR in the 1970s and 80s.

China has denied all accusations and continues to insist that Falun Gong, which they also condemn as an 'evil cult', encourages its members to commit suicide, a charge that Falun Gong denies.

The cases that follow are selected from those compiled in a report published by the Falun Gong Information Center on 27 April 2001, covering cases from September 1999 to April 2001.

**Case I: Zhang Yonghong**, female, 39, from Pingdu City, Shandong Province, a former college-educated accountant in a company forced to resign because she appealed her case.

LOCATION OF INCIDENT: Number Six Hospital Pingdu City (a mental hospital) in Shandong Province.

DATE OF HOSPITALISATION: 6 June 2000.

STATE OF HEALTH: healthy.

HISTORY OF MENTAL ILLNESS: none.

UNIT(S)/PEOPLE RESPONSIBLE FOR MISTREATING PRACTITIONERS: Pingdu local police department; Dr Jing (first name not clear), female, over 40 years old, of Number Six Hospital.

COURSE OF EVENTS: after Zhang and two other practitioners rode bicycles more than 900km to Beijing to appeal their cases, the local police and several of her co-workers escorted her back and held her in an office on the first floor of her workplace. An hour later, a car from the hospital arrived with a man and woman who tried to take Zhang away. When Zhang refused to get into the car, the driver and the nurse grabbed her hair, pulled her to the ground and then pushed her into the car and took her to the hospital. There was no written agreement for her hospitalisation; her incarceration was forced. The doctor responsible for her treatment, Dr Jing, conducted no diagnosis and did not record any result of her examination.

TREATMENT: Zhang was mainly given medicine, injections and sometimes shocks with electric needles. The director of the hospital, the doctor and the police talked with her and asked her to give up Falun Gong, which she refused to do. She was forbidden to read Falun Gong books or practise the exercises. If she did she was tied to the bed and forced to have treatment without her consent. She was not informed of the diagnosis or given any information about the medicine that was being used. At no time was the nature of Zhang's 'illness' revealed to her but, from listening to the doctor's conversation, Zhang learned that they wanted to induce her to forget about Falun Gong by giving her medication.

She was given up to five types of medicine three times a day. In addition, she was given an injection every night. Punishment for refusing medication was that she was held down by male mental patients or hospital security staff and the medicine forcefully administered through her nose. They sometimes used electric needles to punish her. In addition, they increased the amount of medication and forced her to take yet another type of unknown medication.

As a result of the medication, Zhang's thought processes became slow and she lost her memory. Her eyes became inactive and her face yellowish. Her reaction time slowed down and she would often feel sad, unable to control her tears.

She was discharged from hospital after 60 days with no discharge certificate and no follow-up treatments.

Zhang's health was restored three months after she had resumed her practice of Falun Gong.

**Case 2: Li Li,** female, 30, Junior High School education, a former cashier in a grocery store forced to resign because she appealed for Falun Gong.

FAMILY CLASS STATUS: official.

STATE OF HEALTH: healthy.

HISTORY OF MENTAL ILLNESS: none.

UNITS/PEOPLE RESPONSIBLE FOR MISTREATMENT: Pingdu local police department; Dr Jing.

COURSE OF EVENTS: after Li went to Beijing to appeal, the Pingdu local police department told her family members and workplace that three methods could be used to deal with Li's appeal. They could:
1. continue to detain her in her workplace;
2. send her to a labour camp;
3. put her in a mental hospital.
Her workplace said they did not want to retain her any longer since she had been detained so many times. With no choices left, her family members agreed to send Li to a mental hospital. They told her that they were going to take her to the Political Law Committee but sent her to a mental hospital, where she was admitted to Number Six Hospital in Pingdu City on 8 June 2000. Li did not consent to her hospitalisation and there does not appear to have been any written authorisation. There is no evidence of a diagnosis and no written diagnostic report.

TREATMENT: Li was mainly given medicine and injections. She was sometimes shocked with electric needles. The doctors prohibited her from practising the exercises and asked her to give up Falun Gong. In addition, the local police asked her to sign material denouncing Falun Dafa. They promised to let her go if she signed the agreement; if not, they said, she would stay in the hospital indefinitely. She was not informed of her diagnosis or given any information about her medication. No aspects of her illness were explained to her; she was not told of the side effects of the medication.

She was given a handful of drugs each time. Punishment for refusing to take this was that doctors and nurses held her down, grabbing her arms and legs. Some held her head and body, others forcefully administered the medication through her nose. Two people held her down on a bench and tied her arms to the back. They force-fed her one bowl of medicine after another through her nose almost causing her to choke to death. The first time she refused to take the medicine she was tied to a bed and force-fed with an unknown medication.

The medication and injections made her overweight and weak; she felt sleepy all day. She couldn't stop salivating. Li lost control of her arms and legs and stumbled when she went to the bathroom. She could not stand up once she squatted. She couldn't open her mouth when she wanted to eat.

After being discharged from hospital 123 days later, she suffered from terrible headaches and could not sleep since her eyes remained open all night. Her health was gradually restored once she resumed her practice.

**Case 3:** practitioners from Wuhan, Hubei Province, including **Xu Jiangong**, a retired university worker.

Xu Jiangong went to Beijing to appeal his case in early December 2000. After he had been arrested and brought back to Wuhan, Xu was detained and then transferred to Yatai Mental Hospital at Wuhan University. More than two months later he was still being held there.

**Case 4: Huang Lilan**, female, 68, a retired official from the Bureau of Meteorology, Guangdong Province.

LOCATION: a mental hospital in the Baiyun District, Guangzhou City, Guangdong Province.

UNIT(S)/PEOPLE RESPONSIBLE FOR MISTREATMENT: Xu Jubo and Fang Guohui, officers from the first section of the Guangzhou Dongshan Police Substation.

COURSE OF EVENTS: Huang Lilan is an honest person who abides by the law and behaves herself in all aspects of her life. Because she practised Falun Gong, however, a group of plain-clothes police officers broke into her home on 11 October 2000 to take her to the local police department. Thinking the police would follow the correct legal procedure to arrest her, Huang asked the police to show their identification. The police said they had nothing to show her. Instead, they forced her down the stairs and pushed her into a police vehicle waiting for them. Huang was sent to the mental hospital in Baiyun District. The only person present during her arrest was her house guest. The police sent her to the mental institution without informing her family. In order to keep the news from the public, the police detained Huang's guest in a place near Luogang Station in Guangzhou; the guest was forced to pay 600 yuan (approx. US$75) to be released.

Police told the hospital officials that Huang's personality became extremely abnormal after she began practising Falun Gong. She was sent there, they said, for disturbing the stability of society. Huang Lilan was badly tortured in the mental hospital. Her first day in the hospital, Huang was tied in a chair for the whole afternoon. She was also given unspecified medication by injection. The hospital began to give her large amounts of medicine twice a day. Huang said the medicine made her constantly sleepy at the beginning, and she felt dazed. After the injections, Huang's appearance began to deteriorate. As the injections continued, the side effects became increasingly strong; her mouth was dry and had very little saliva. To keep her mouth wet, she drank large amounts of water that caused her abdomen to swell. Even then, Huang's mouth was still too dry. Because of the large amount of water she was drinking, Huang needed to use the lavatory frequently throughout the night. In addition, Huang began to develop insomnia. She understood that she could die and the public would never know the truth.

Huang's workplace talked with the local police department and, with their help, Huang was able to go home. At that time, she had been in the hospital for three months.

Meanwhile, the police had ransacked Huang's home and removed any paper with the three characters of Falun Gong written on it. They also associated Huang's telephone and yellow clothes with Falun Gong and removed them. ❏

***Danny Schechter** is executive editor of Globalvision's Mediachannel.org and the author of the recently published* News Dissector *(Akashic Books) and the revised edition of* Falun Gong's Challenge to China. *The full version of Falun Gong practitioners who are detained and tortured in mental hospitals is available at www.faluninfo.net*

FRANÇOIS VINSOT

# Back to normal

**What future for the victims and perpetrators alike after the madness of genocide?**

They were seven years old at the beginning of this year, the children who were born as a result of rape during the genocide. If they were really lucky they just ended up as ordinary orphans.

But then there are the others, the children who, during the genocide, became killers. Those that I met soon afterwards grinned as they argued about who had killed the most. If they were really lucky they just ended up as normal adults.

He was eight when the militiamen arrived, but only six when his father set off every morning to join in the massacre. His mother joined in as well. Then some youngsters turned up and they were killing people too, so they took him with them. He tagged along and joined in the killing with them. They helped him, especially at the beginning, because he wasn't strong enough and he hadn't got the knack of it, but drinking with the others made it easier. He wasn't scared: at first he wanted to get it over with as quickly as possible, but then it went on for several days. They left him behind quite often because he couldn't keep up or they gave the machetes to the most expert killers; but when they were setting fire to churches he was able to help. He knew the people they were killing: he recognised the schoolteacher and the other pupils, but there was so much noise that he couldn't hear what was going on and the others were going so fast. When they had finished with the village, the militiamen left and took him with them. While they were fleeing he was recaptured. What happened was that their truck came off the road and the pursuers caught up with them. He was nearly killed but they decided to keep him, to make an example of him, you can say, and they tied him up in the back of the van. People in his village recognised him and there was an organisation that was trying to keep the children separate from the others to try to re-educate them; then the army took charge of him.

*Kabgayi, Rwanda, 1994: 30,000 Tutsis are herded into a concentration camp.*
*Credit: Gilles Peress / Magnum Photos*

Those victims who escaped death carry on as best they can, often not very well. What they say today is what they said yesterday and what they will go on saying: for them time came to a halt and they can find no peace of mind. They complain that they have been abandoned. They are the ones who have to face all the grievances, sometimes compassion, sometimes others' shame for what they have done. At first sight they seem to be enclosed in a silence so profound it's frightening. Their silences seem fathomless; but then, sometimes, just a word, just a look, just a few moments' wait will turn a victim into an eyewitness. In a feeble but clear monotone they will tell you, as they stare at the ground, how they escaped the worst fate; they're alive, they're lucky; that's what they say. And one of the first things they tell you is that they are one of those whom death refused. Then they describe what they have witnessed, acts of unbearable horror.

They also claim that they are ready to forgive but no one has come and asked them to.

About seven years ago, Catherine was left for dead in a heap of corpses. She kept perfectly still until everything was quiet again. Since then, she's lived on as an officially declared survivor, one of those statistically estimated one in ten survivors among Tutsis known to have been living in villages in Rwanda when the militiamen came to murder them systematically.

She is the one who survived for every nine who died of those who witnessed the genocide, who saw the massacres. She is one of those for whom this word genocide must carry all the weight of their hope to be, if not avenged, then at least recognised for what they are. Catherine knows what she is saying, knows what she wants. During the day, she's OK; as long as she's busy, it's OK. The night-time is hardest but she's coping and, anyway, she's got so much to do, so many things to fight against; she hasn't time to weaken. The genocide has taken over her life: there's no end to it and there never will be. She has no spare time: 'I don't want any. I don't even want to try to live normally. What's normal, anyway? What's left that's normal about life after genocide? Even if I wanted to I couldn't.

'Sometimes, with a few close friends, we've got together for a few drinks, tried to talk about something else, but we never managed it. After a minute or two there was always somebody who would start to reminisce, who'd come up with some little detail, something he'd never said before, something he'd forgotten or that he never thought about, and off we'd go. Everyone would have something to say; even the silences were different then, and so, after a few times, we gave up saying that we'd been going to do something else. We meet, we know we're going to talk about it, and we do. Sometimes we don't even really need to use words, or else we talk about the present, but the present is going to take us back to the genocide; it's going to make us talk about children with no parents, about children who caught AIDS from the man who raped them.'

Officially, Catherine escaped the worst fate but, as one of her acquaintances says, it was pretty close, and anyway, things can still take a turn for the worse; it all depends on aid, and aid is not always delivered as it should be. Catherine can read and write, so it was easy for her to find work. She can speak several languages too and can use this to help other victims. Catherine is staggering under the weight of her responsibilities and the decisions she has to take: that is her salvation. Gradually she raises her head and, over her glass of lemonade – she drinks no alcohol – I look her in the eyes and she meets my gaze and I see what I've forgotten. It's a blend of emotion and pain, a fearful rigidity of expression. Raking through my memory, I recall the facial expressions of some of the patients in a psychiatric hospital in Paris after years of treatment and being fed powerful sedatives; patients who have been classified as a danger to themselves and to others and so are kept locked up. This woman, even though she is in the open air, is locked up like them: night and day she struggles to forget and to remember, to bear witness and to try to get on with her life. The victims are shut away behind invisible walls and they have no means of escape: they have no future. It has been stolen away because no one kept it safe. ❑

***François Vinsot*** *was BBC Central Africa correspondent*
*Translated by Michael Routledge*

**VLADIMIR BUKOVSKY**

# Roll up! Roll up!

**Vladimir Bukovsky was first arrested in 1963 for the possession of anti-Soviet literature and interned in a special psychiatric hospital for 14 months. In 1965, he was rearrested for co-organising a Moscow human rights demonstration in defence of writers Andrei Sinyavsky and Yulii Daniel, and sent to a series of mental hospitals. Released in 1966, he was detained again in 1967 for organising a demonstration and sentenced to three years in a labour camp. His fourth arrest in 1971 led to a further seven years in prison and five years' exile. In December 1976, Bukovsky was released in exchange for the Chilean prisoner Louis Karvallan. He was interviewed by Irena Maryniak**

*You were twice subjected to compulsory psychiatric treatment in the 1960s, first in Leningrad's Special Psychiatric Hospital and then in 'ordinary' psychiatric clinics. What were the conditions of your internment and how did it affect you?*

I was young and curious so I wasn't seriously intimidated, even though the chances of never getting out of hospital were high. But I didn't know that then. The special hospital was an overtly penal institution. The orderlies were criminals who had been designated to do time there. As far as they were concerned there were no rules, anything went. They could steal, beat or kill. The authorities never punished them, they put everything down to the patient's condition. If an orderly killed a patient, it was the patient who was to blame.

Unofficial forms of punishment included injections and 'roll-ups'. I never had a 'roll-up', but it was an ugly thing to watch. They'd wrap wet canvas strips around the patient which shrank when they dried so that the victim lost consciousness. Then a nurse would come and loosen them. This could go on for any length of time. They called it 'restraint'; in fact it was torture. Some people suffocated and died. A young political

I knew, Tolik Belyaev, was given this treatment because he had been reading after lights out.

The punishment drug aminazine was widely used – in the early 1960s they had little else. They also gave sulfazine which was a solution of sulphur in peach oil, injected into muscle. It caused an abscess, high temperature and intense pain. In one section they gave insulin shocks, but we were spared those. Electric shocks were introduced later. They didn't have the technology then. Subsequently, the range of available equipment grew much wider.

*Cambridge, 2001: Vladimir Bukovsky. Credit: Irena Maryniak*

They also administered the drug haloperidol. The idea was that it lowered the severity of the psychotic state by affecting neurotransmitters. An excess of dopamine induces a severe psychotic state and a low level of dopamine causes Parkinson's disease. They gave high doses of haloperidol to lower the dopamine level, and people got symptoms of Parkinson's.

All this made everything that came after seem much easier. When I was imprisoned, I found it quite simple to tolerate punishments like solitary confinement, cold or hunger. They didn't touch me. In prison you had colleagues, cell mates and limited rights. We used what we had. We went on hunger strike . . . In psychiatric hospitals there was nothing.

*What about friendship?*

We had that, but in those conditions there wasn't much we could do for one another. It was like a mischievous trick. You'd been diagnosed mentally ill. You were no longer responsible for yourself. You had no rights, even theoretically. I was fortunate in that our doctor was well over 70 and very experienced. If an orderly said that a patient had woken up and attacked him, he'd say: 'I've been working here 40 years and I've never seen a patient get aggressive. What did you do to him? You must have done something . . .' So in his section they never touched anyone. There was also a woman doctor who secretly helped us. She still lives in St Petersburg. She was an 'ascetic' in the Russian Orthodox sense of the word. She worked in mental hospitals to help people. She wouldn't let them punish us and helped to get us discharged. I kept in touch with her quietly after my release. She went on giving help, and was eventually caught and sacked. I lost sight of her after that. In 1991, I found her again, destitute, penniless, with no teeth. She was in a horrifying state. Her former patients now send her money from abroad.

*What was the relationship between real patients and political prisoners?*

We made a joke of them. Over half of the patients in our section were healthy – it was considered a 'soft' section. In 'hard' sections the number of sane patients was very small. About 10% of all patients in Soviet psychiatric hospitals were politicals. A hospital might have 1,000 patients, of whom 100 would be mentally sound. Many of the others were multiple murderers; they might have killed in desperate circumstances.

*How did ordinary and special hospitals differ?*

People were sent to special hospitals following a court decision. There'd be a trial and if the investigation found you mentally irresponsible you'd be sent for compulsory treatment. But you could end up in an ordinary hospital on the basis of complaints from neighbours or relatives. People turned up fortuitously. Some had tried to commit suicide, others were merely thought to have attempted suicide. This was categorised as 'dangerous'. A potential suicide was dangerous to himself. Anyone who attempted suicide and survived had to submit to psychiatric treatment. You weren't supposed to commit suicide – it was seen as deviancy.

In ordinary hospitals the percentage of people who were genuinely ill was high and most came of their own accord, through a doctor. The different sections were graded. There was a ward for the chronically ill who had been there for decades. There were also wards for alcoholics and drug addicts.

*In* To Build a Castle *you describe how it felt to leave hospital and go back into what we think of as 'normal' life . . .*

It was a reaction many people experience after release from imprisonment of any kind. The process of reintegration is intensely difficult. Getting used to prison is far easier. In extreme conditions you discover a greater capacity to adapt and react quickly. When you're freed you expect normality and there's no such thing. In prison you idealise life and freedom, it's like nostalgia for an imagined world. The mind embellishes what it wants. When you're released you perceive that things are quite different. It's a passage from one universe into another. You need time – and when you're set free you don't get that. There's a life to live, there's work, and people don't realise what's happening to you. It's as if you'd changed your skin. You walk around bare-skinned, highly sensitised. As a prisoner you're subjected to sensory deprivation. I remember the first thing that struck me when I came out of prison was colour. You're out of the habit of seeing colour – nothing coloured is permitted in prison. Even the strength of colour is painful and demands reintegration.

It's also a time of rejection. You want solitude but you're in work, there's family, friends keep dropping in. You don't want that. If you're just out of psychiatric hospital it's twice as bad because of the psychological tension there. You're constantly wondering if you're normal. Even though you know you were diagnosed for political reasons you still watch yourself. Perhaps I am mad? Those big nobs in white coats with diplomas and professorial status decided I was. There must be something wrong. You keep analysing yourself, comparing yourself with others. It's an additional burden.

*You must have thought a lot about mental illness and what it is.*

I saw too much of it. I had the feeling it was like a technical fault, an engine running after something's seized up, or one of those old gramophone records that goes into permanent replay. No one has really understood schizophrenia yet.

*It's ironic that Professor Andrei Snezhnevsky of Moscow's Serbsky Institute apparently thought he did. The story of the struggle between the two Soviet schools of psychiatry, 'Moscow' and 'Leningrad', is well known. How different were they?*

Snezhnevsky dreamed up a new form of schizophrenia: 'sluggish' or 'creeping' schizophrenia. The idea was that schizophrenia can begin in early childhood as a result of psychological shock that evolves into clear symptoms only years later. The problem with this was that there were no objective criteria. Living in the Soviet system it was virtually impossible not to be traumatised. Snezhnevsky saw potential schizophrenia from early childhood in everyone. You could show him anybody and he'd say 'schizophrenic'. I knew him quite well and I think part of him really believed it. But it was very convenient for the KGB. In any other country his views would have raised a laugh or prompted debate; he would never have dominated his field. Here he was useful without realising it. I don't think he understood it for a very long time, only towards the end. If they wanted someone diagnosed as mentally ill he'd do it. It was as simple as that.

The Leningrad school was more traditional. They weren't intellectual giants, they were professional psychiatrists and didn't hold with all this nonsense. So if you were diagnosed with schizophrenia in Moscow and taken to Leningrad, they'd often say you were fine. 'And as to the future – who knows? We don't have a crystal ball.' Sluchevsky – the leading Leningrad psychiatrist – was an old, highly intelligent and educated man. He regarded Snezhnevsky's theory as absurd. In the 1960s, he was very influential in Leningrad, so if they brought him a patient from Snezhnevsky he'd delete him from the list without a second look. This discrepancy explains a whole series of cases. The dissident Marxist-Leninist Pyotr Grigorenko was diagnosed twice as healthy and then taken to Moscow and diagnosed as mentally ill. It was a time of covert attack and counter-attack between these two schools of thought. But it ended soon enough because, with the support of the authorities, the Moscow school prevailed and became obligatory.

*In the 1970s you collaborated with Semyon Gulman on* A Manual on Psychiatry for Political Dissidents, *which gave advice on how to deal with internment in psychiatric hospitals. Did you follow your own recommendations?*

Not always. Some of it came from experience, some was put in because Slava (Semyon) felt, as a psychiatrist, that it was necessary. Some of it was intentionally malicious. A sort of joke to demonstrate the paradoxes of the situation. But it was useful; the advice is sound. People like us were quite unprepared.

*You suggested that people should, if they had to, retract their beliefs.*

We had to explain that there would be a dilemma. I can't advise anyone

*Sergei Chepik,* The Madhouse, *1984–87. Credit: The Bridgeman Art Library*

to deny their own views. I didn't do it myself. But people needed to know that there might be a moment when they would have to choose. They needed to be ready. After that, it was up to them.

I was lucky. The doctor said to me: 'I expect you're pretending. I can't see any symptoms. How did you get in here?' 'You'd better ask them,' I said. He kept demonstrating that I was mentally sound, which wasn't what the authorities wanted to hear. So, in the end, they compromised: they said my condition had 'improved' and retired the doctor.

I thought about what I might do, of course. You didn't usually get discharged from psychiatric hospitals unless you admitted your 'mistakes'. In the event, it proved simpler for them to let me out. But many others had to do it. If you had a family you were very vulnerable. But I had comparatively few Achilles heels and never had to make the choice.

*What is happening in Russian psychiatry now?*

Systematic abuse is over. The authorities have no interest in it, there's no demand. But occasional abuses occur, even in St Petersburg, and religious groups are sometimes still exposed. I recently received reports about regional authorities persecuting religious communes. The Moscow Patriarchate is often involved because it doesn't want competition, or needs a new church building, whatever. Sectarians are diagnosed as psychiatrically ill. It's convenient. They're taken away and there's more room for manoeuvre. These are localised cases. It doesn't happen in Moscow. There it's simpler and cheaper to kill people than to imprison or hospitalise them. Moscow is indifferent. This isn't the age of Andropov. Psychiatric diagnosis implies a complicated process once controlled by the Party and the system. Today there are voluntary groups of psychiatrists who monitor what's happening. So incidents are publicised and a system can't be built up. You get isolated cases, but not a system. I suppose that can be considered a success.

*What do you make of the Russian Psychiatric Society's estimate in April that, today, a third of Russians suffer psychological disorders?*

Russia was traumatised by 73 years of communism followed by a sudden transition to capitalism. It's hard to find anyone who hasn't been personally traumatised. The new generation may be normal – time will tell. Many older people are psychologically broken. The experience of totalitarianism was immensely hard: total dependence and uncertainty, the arbitrary abuse of power. You had no idea what the authorities would do with you tomorrow. Even the existence of two conflicting channels of information was a trauma. I say that as a neurophysiologist. Radio, television, the press said one thing; life showed you something very different. It's a classic way of inducing trauma or neurosis. The way Russians escaped this discrepancy was to get drunk or tell jokes. There was a well-known anecdote in which a man goes into a hospital and demands to see an ear-eye specialist. They say to him: 'There's no such thing. There are ear, nose and throat specialists and eye specialists. What do you need an ear-eye specialist for?' 'I've got this problem, I don't see what I hear and I can't hear what I see.' Thousands of anecdotes were told in response to the lies people were fed, and to the internal, psychological conflicts this provoked. And then there was the hopelessness, the impossibility of getting out. It was deeply damaging for all those who lived through it.

Today people are still highly suspicious of one another. No one believes what they hear. There's always something behind it, and something beyond that – layer after layer like a matrioshka doll. That's how life was arranged. No one will ever do anything simply. You have to find a way round, take a detour. They talk about creating 'market relations'. You won't get 'market relations' in Russia so long as people don't have a direct relationship with things, let alone with each other.

*Shortly before your arrest, Nikita Khrushchev declared that everyone was happy with the communist system and that those who expressed dissatisfaction had to be mentally ill. Do you think the view that if you're in conflict with society you must be mad still carries weight in Russia?*

Today's generation, the people we call 'Generation X', believe in total conformism. It's all they have: a reaction against the excessive idealism of their fathers and grandfathers. We were too idealistic and our children are conformists. There are similarities in Britain, especially now. It's even worse in the US. American society has a mob psychology. Ask for salt and you're an enemy of the people. In the West people don't know what historical experience has taught us in Russia: that conformism is dangerous. It's the foundation of totalitarianism. It was the same in Nazi Germany. Brecht writes a lot about conformism. There are always small groups of fanatics, but if a society is healthy enough it won't let them impose their madness on the rest. But a society predisposed to conformism will succumb.

*How do you explain Russia's recent restoration of some aspects of Soviet rule?*

They didn't carry things through and break the system. In 1991, we suggested putting communism on trial to draw a line under it once and for all. It didn't need to be a trial of personalities as in the Czech Republic – individuals could be left alone. We should have tried the system so that people understood to what extent they had participated in it. It was essential, and they didn't do it. The *nomenklatura* hung on to power and subsequently took the offensive. The former KGB has taken control of the mafias, the administration, the legislative apparatus, everything. This is as dangerous for the country as it is for its neighbours. But it can't last. People didn't understand why the Soviet Union fell. They didn't see that it was inevitable. They thought the CIA or a criminal conspiracy had undermined the state. They blamed Yakovlev or Gorbachev even though these were the people who wanted to maintain the regime, not destroy it. The people in power today are revanchists. They are trying to create a partial image of the Soviet Union without understanding that it can't be done. They could cripple the lives of yet another generation. Russia has minimal chances of recovery as it is. The chance that it will simply die is very high. If you break another generation the country will disintegrate.

*You haven't been given a visa to travel to Russia since 1996. Why?*

Since the mid 1990s, the ex-KGB has been taking over key positions in the administration and preparing to get their man to the top. They're still 'myth-making' in a way. They imagine that people like me can do something. They're terrified, they always were. I've seen documents showing that they always thought there were more of us, that we were stronger, that there was something behind all this when there wasn't. In their eyes we became giants. In fact there's nothing I can do. I don't even try. They recently denied Alexander Ginsburg a visa. He isn't well, he's only got one lung. What have they got to fear? He goes over to buy books. But they didn't let him in just in case. We're trapped in myths.

*And paranoia?*

There's never been any shortage of that. They spent decades hunting enemies of the people. They invented them. How could you avoid going mad? It was the way the system worked. Look at the resolutions passed by the Politburo, the minutes of the meetings they held. It was a nuthouse. I discuss them in my latest book, *Judgement in Moscow*. In 1992, I was invited to testify at the trial the Russian Supreme Court conducted against the Party. I was given access to the Central Committee archives and to some of the Politburo archives. I scanned many of the documents and published a book.

The Politbiuro would be discussing a problem and they'd say world imperialism was behind the Solidarity movement, for example. It's impossible to tell whether they believed what they're saying. On the one hand they had to talk as ideology decreed, on the other they needed to be effective. They understood that it was an ideology, they weren't naive. But the power of inertia, the habit of believing the doctrine had done its work. They were fighting shadows. It was a cast of mind in which they had grown up. And even if you were cynical enough to understand how relative it all was, it didn't mean that you were free in relation to it.

I examined records of meetings on the invasion of Afghanistan, for instance. The Politburo spent six months persuading themselves not to send troops: everyone understood that it was a mistake. Then they went anyway. It was Moscow's decision to destabilise the country by removing the Shah. They put the former prime minister, Mohammad Daud, in charge. Five years later, the 1978 April Revolution took place. But in Moscow they seemed to be incapable of understanding the level of instability they were provoking. I found a remarkable dialogue between Soviet Premier Aleksei Kosygin and the PDPA Khalqi faction leader Nur Mohammad Taraki. When the 1979 crisis began, Kosygin telephoned Taraki and they had a conversation reminiscent of something out of Tom Stoppard. A division had rebelled and it looked like the end. 'Why aren't you arming the workers?' Kosygin said in time-honoured Marxist style. 'We don't have workers, we have artisans.' 'What about the students?' 'We don't have them, we have schoolchildren.' 'Where are the officers we trained for you?' (Afghan officers were trained in Russian academies.) 'Oh, they've all gone over to the enemy.' The Soviet Union had provoked revolution in a pre-industrial country which was, by any Marxist criteria, totally unsuitable. And then it wondered why it couldn't stay in power. Marxist ideology was spectacularly unrelated to reality.

*And today's paranoia is a vestige of this . . . But isn't it understandable that Russians might be genuinely concerned about how richer and more powerful countries could treat them?*

They're more cynical than that. People are filling their pockets and the KGB has degenerated into a Bond-style crime cartel, a 'Spectre'. Its concerns are more banal than ever: control and money. In that sense the paranoia has subsided but the mindset is still there: suspicion, mistrust. In the end it can only turn against itself. ❏

***Vladimir Bukovsky**, dissident and neuroscientist, was interviewed in Cambridge by Irena Maryniak, who also translated the interview*

# Poems from the Arsenal

In the mid-1970s, I met the dissident Victor Fainberg and the psychiatrist who saved his life, Marina Voikhanskaya, at an *Index on Censorahip* party. CAPA (Campaign Against Psychiatric Abuse), or 'Capitolina Ivanovna' as Victor dubbed it, was formed and I was happy to work alongside Victor and Marina, the late David Markham, the redoubtable Max Gammon, William Shawcross, the late George Theiner of *Index*, James Thackara, Tom Stoppard, Eric Avebury and others. We campaigned for the mathematician Leonid Plyushch, Anatoly Sharansky, the poet Zinovy Krasivsky and many more; and, of course, Vladimir Bukovsky.

Victor told me that in Leningrad Arsenal Mental Prison Hospital he had been given a little book, bound in Elastoplast, containing about 100 poems written in a regular but difficult-to-decipher hand. The man who gave it to him was a 'thief of honour': he claimed to have written out the poems clandestinely in one night, and said they were by many hands.

Victor thought the poems might be by a poet or poets who were 'genuinely mentally ill' (which he was not). I translated them all in the 1970s – they were published in a number of places and I tend to think they are by one hand. The Arsenal was used as a punitive psychiatric hospital for dissidents. Massive doses of Aminazine (chlorpromazine/largactil), known as 'the liquid cosh', and the punishment drug Sulfazine (which violently raises the temperature) were regularly used.

The images are strong, with a suppressed anger, escapism; the love poems are anguished. Their mysterious origin adds to their effect and they are representative of a significant segment of life and poetry in Russia in the 1970s. The feelings of incarceration are authentic. If the Unknown Mental Patient has survived somewhere in Russia or wherever, I would like to return his poems, which he gave to Victor in the hospital when he knew he might get to the West.

***Richard McKane***

ROUND THE GLOBE the train of dust,
round the globe the dust of mirages,
the shadow on the globe turns black,
the globe in the shadows of lost days.

The emerald sky has stretched out
over the planet of maddened shadows,
towers have thrust their voice into the heavens,
and exploded in the chaos of steps.

The chain swings on the wall,
as a pendulum of an antique clock,
and I am in my mad fire
of voices half-sane as the day.

IN THE HUMPBACKED Arbat streets
an alien man got lost.
Much seems to him to be strange,
in the noisy autumn peoplessness.

The street lamps and shop windows in a torrent
of rusty light splash in the dark,
and the windows of people's thoughts, of beasts
look at the autumn night.

Noisy crowds rush past
in the howling of the cold wind,
and the desert of the autumn peoplessness
is bloodsucked with the hubbub of greed.

In the humpbacked backstreets of the Arbat
an alien man got lost –
he is like a beast captive in a bestiary,
and does not find the way out of the cage.

ON THE GLOOMY horizon of the cell
no hope of the star's shining
only a sun yellow like sardines
and the dull waiting – when?

Waiting surrounded the cell
like a ghost of dead minutes,
it's the years of pains and distances,
I walk in delirium I fly in delirium.

On the gloomy horizon of the cell
no hope of the stars leaving a track.
Unhappiness menacingly rules in the cell,
with a station change for the other world or this world.

This cell – the delirium of my fantasy,
if I am a ghost among people,
this cell is a door to the expanse of the *Odyssey*,
the voices and wishes of the people.

THE MOLE STRAYED into the green of the waves,
the peals of thunder shook the sieve.
The smoke of distant gullies arose
and floated like a transparent shroud.

The mist whirled into a column
under the greying, tired sky,
and the caravan got lost in the expanses,
and the paths were sprinkled with poems.

The song goes away with the caravan,
the song of the sky of the sands of the oceans.
Lava of snow flows from the mountains
and smashes to pieces like the roaring of a tiger.

DON'T COME to me,
it's difficult for me to talk with you.
I cannot love you,
and it's not in me to give you
the breath of joy.
Don't come to me.

The years have closed tight shut,
in the abyss of terrible distances
the flamelets of desires have died.
You have become a memory deceived,
you are somewhere there,
the years have closed tight shut.

Don't come to me.
I shall not return to your crystal world.
You are the distant echo of a song.
You were for me, but became
that which one loses without finding.
Don't come to me.

AN UNKNOWN track flares
on the wiped-out line on the horizon,
(someone walked over the earth in the rain
and is continuing on his way in heaven).

The horizon
is the frontier of two worlds,
it's difficult to cross it today,
and tomorrow it will be absolutely impossible.

So one has to gather all one's strength
on the frontier of tomorrow and today,
and to go away leaving between them
all that these two countries,
today – the present,
and today – the tomorrow of the past,
compared you to,
throughout the many ages.

AGAIN THE DROPS of red-brown liqueur,
again
raindrops
beat
on the windowpanes.
Again the mountain air of Daghestan.
Again your echo
close
beside me.

. . .

But when will these
disappear at last:
this table
this pain
this abyss
and the constellations of arms
intertwine
and the mists of parting
will melt
and I will see the whirlpools of your eyes
and fill them with the sun's happiness?

Raindrops again
Beat on the windowpanes.
Come to me
my distant happiness. ❑

*Introduction and translations by* ***Richard McKane****, who is a writer, translator and poet. The poems have been published in* Gnosis, *under the pseudonym of Ivan Bulyzhnikov (John Cobblestone), in* The Month, *in* Beyond Bedlam *(Anvil Press) and a dozen in his* Poet for Poet *(Hearing Eye)*

More articles, Index archive material and link ⇨ www.indexoncensorship.org/madness
Put your opinion online ⇨ www.indexoncensorship.org/comment
Email the editor ⇨ editor@indexonline.org

STRIP SEARCH
by Martin Rowson
1.
YOU DON'T HAVE TO BE MAD TO WORK HERE... BUT IT HELPS!
IN CASE OF

4
YOU DON'T HAVE TO BE MAD TO WORK HERE... BUT IT HELPS!

6
UNHCR
YOU DON'T HAVE TO BE MAD TO WORK HERE... BUT IT HELPS!

2
YOU DON'T HA
TO BE MAD T
WORK HERE
BUT IT
HELPS!

3

5
YOU DON'T HAVE
TO BE MAD TO
WORK HERE –
BUT IT HELPS

You didn't have to have been mad
to have lived there... But it helped!
© Martin Rowson 01

## FLASHPOINT

# Under cover of elsewhere

Eritrea's entire independent press was closed down on 18 September as Eritrean President Issayas Afewerki moved against the self-styled 'Reformers' group of dissidents within his own ruling People's Front for Democracy and Justice (PFDJ).

The closures – involving eight papers – were followed less than a week later by the arrest of nine of their journalists. Two reporters still on national service have been scooped up from the army and sent to work on a state-owned gold mine. Others reported 'calls' from the *Hagerawi Dehn'net* (National Security) force and were expecting arrest as *Index* went to press. Others have headed for safety in Sudan and Europe. The crisis follows growing frustration with the legacy of the country's disastrous two-year border war with Ethiopia and a split between Afewerki and many of the PFDJ's senior membership. The increasingly self-isolated Afewerki was refusing to call meetings of the Party Central Committee or recall the National Assembly. Neither body had met for 20 months before the swoop.

'It was a simple way to avoid reality,' said an Eritrean political analyst in London. 'He simply bided his time.' The trigger appeared to be September's attacks in the US. 'President Afewerki's government is apparently trying to use the world's current preoccupation with other events to escalate its repressive campaign against dissidents,' said Suliman Baldo, senior researcher in the Africa division of Human Rights Watch. The centrepiece of the operation was the arrest of 11 of the so-called group of 15 (G-15) dissidents – all members of the ruling PFDJ Central Council as well as the Eritrean National Assembly. All were signatories to a widely publicised letter to PFDJ members in May, criticising Afewerki's 'illegal and unconstitutional' style of rule and demanding that the National Assembly be convened.

A follow-up letter in August urged unity but reiterated the problems that, by implication, Afewerki's rule had failed to tackle: post-war reconstruction, unemployment, the abuse of women in the army and the plight of poor unpaid soldiers on the front line with Ethiopia. The 'reformers' had made wide use of the independent press to publicise their views. Among the arrested were former interior minister Mahmud Ahmed Sheriffo and two former foreign ministers Petros Solomon and Haile Woldensae. Also arrested were former senior official in the finance ministry General Estifanos Seyoum, former information minister Berahi Gebreselassie, former labour minister General Ogbe Abraham and former military commander Berhane Ghebre Eghzabiher.

Of the others, one has recanted and the other three are in the US. Amnesty International fears they could be arrested if they returned to Eritrea. 'Those arrested may be prisoners of conscience detained solely for the peaceful expression of their political concerns,' Amnesty added. The jailed 11 are thought to be in detention at a government centre outside Asmara. They have not been given access to their families or lawyers and there are growing fears about their safety.

The May letter represented the first sign of top-level dissidence since Afewerki became head of state on Eritrea's independence from Ethiopia in 1991 after a war spanning three decades. The December 2000 peace deal between Eritrea and Ethiopia followed, officially ending a conflict ignited in 1998.

The clampdown also raised tensions among students at Asmara University, focal point for public protest all summer. University students' union president Semere Kesete was arrested and 'disappeared' by security forces on 31 July. The students are expected to spend the summer break at community work camps in exchange for deferment of their military service. The conditions at the labour camps are appalling, they say. Kesete was arrested a day after publicly announcing that students would not join the summer camps unless conditions were improved. Some 400 students who tried to protest against his arrest on 10 August were immediately sent to one of the worst such camps, in Wia, near the Red Sea port of Massawa. Another 1,700 students at the university, the only one in the country, voluntarily joined them at Wia. According to Human Rights Watch, at least two, maybe more, had died in the camps from heat exhaustion by late September. Officials had offered students the chance to leave the camp and return to Asmara to register for the new academic year at the end of August, but withdrew the offer when the students literally turned their backs on the university's president when he visited the camp. Police have also used batons on protesting parents demanding the return of their children from the camps. ❑

***Rohan Jayasekera***

*A censorship chronicle incorporating information from the American Association for the Advancement of Science Human Rights Action Network (AAASHRAN), Amnesty International (AI), Article 19 (A19), Alliance of Independent Journalists (AJI), the BBC Monitoring Service Summary of World Broadcasts (SWB), Centre for Journalism in Extreme Situations (CJES), the Committee to Protect Journalists (CPJ), Canadian Journalists for Free Expression (CJFE), Glasnost Defence Foundation (GDF), Information Centre of Human Rights & Democracy Movements in China (ICHRDMC), Instituto de Prensa y Sociedad (IPYS), The UN's Integrated Regional Information Network (IRIN), the Inter-American Press Association (IAPA), the International Federation of Journalists (IFJ/FIP), Human Rights Watch (HRW), the Media Institute of Southern Africa (MISA), Network for the Defence of Independent Media in Africa (NDIMA), International PEN (PEN), Open Media Research Institute (OMRI), Pacific Islands News Association (PINA), Radio Free Europe/Radio Liberty (RFE/RL), Reporters Sans Frontières (RSF), the World Association of Community Broadcasters (AMARC), World Association of Newspapers (WAN), the World Organisation Against Torture (OMCT) and other sources including the International Freedom of Expression eXchange (IFEX).*

## AFGHANISTAN

Reporters in Kabul to cover the 4 September opening of a trial of eight foreign aid workers accused by the ruling Taliban militia of preaching Christianity were first told by Taliban foreign minister Wakil Ahmed Muttawaki that the trial would be held in open court, but finally were given access for the first day only. On 9 September reporters were prevented from leaving their hotel and their rooms were searched. The aid workers and 16 Afghan colleagues worked for the German-based charity Shelter Now International. In July, the Taliban announced that the penalty for a foreigner suspected of proselytising was jail and expulsion. Afghans who preach or convert to a religion other than Islam face the death penalty. (CPJ)

The Pakistan-based press agency *Afghan Islamic Press* announced on 13 July that the Taliban had ordered the banning of the Internet in territories under their control. According to the foreign affairs minister, the ban will prevent access to 'vulgar, immoral and anti-Islamic' content. Only a few Afghans and foreigners working for international organisations have Internet access through Pakistani phone lines. (RSF)

## ALGERIA

Algeria's revisions to the defamation law under its Penal Code were signed into force on 27 June 2001. New clauses include Article 144 (b), threatening three to 12 months in jail and/or a fine up to US$3,200 for 'any person who offends the president of the republic, through the use of an offensive, insulting or defamatory term, be it in written form or drawings, be it orally, in an image, or via electronic or other support.' Article 144 (b)(1) specifies that proceedings will be launched against the publishers and the publishing company as well as the author. Article 144 (b)(2) provides for three to five years in jail and/or a fine of up to $1,300 to those who 'insult the Prophet and the messengers of God or denigrates the dogma or precepts of Islam.' And Article 146 also sets fines and sentences for insulting parliament, the army or any 'constituent body or other public institution'. (CADLP)

On 10 July **Faouzia Ababsa**, editor of the daily *L'Authentique* was convicted *in absentia* of defamation. Ababsa had testified in the case four months earlier but did not know about the new hearing or the verdict until she read about her suspended six-month prison sentence in the press the next day. The case involved accusations made by the journalist in a May 2000 article against the president of a business association. (CPJ)

## ANGOLA

On 8 July **Alegria Gustavo**, a journalist for the local branch of *Rádio Nacional de Angola* (RNA) in the central province of Huambo was shot dead by provincial vice-administrator Matias Kassoma after leaving a party. The journalist's friends then attacked Kassoma in reprisal for the murder and left him in critical condition. (MISA, IFEX)

Angolan journalist **Gilberto Neto** of the independent weekly newspaper *Folha 8* was arrested on 7 July, along with foreign researcher **Philippe**

**Lebillon**, after travelling to Malange in northern Angola without the authorisation of local governor Flavio Fernandes. On 18 August Neto was barred from travelling to London to attend a Reuters Foundation course on business reporting and his passport confiscated. (CPJ, IFEX, MISA)

Angola's Catholic Rádio Ecclésia (*Index* 3/2001) suspended its 8–9 July news services in response to what it called a 'defamation campaign' by the state-owned daily *Jornal de Angola*. The paper had singled out the its coverage of fatal forced removals from the town of Boavista. The clashes served as a pretext for Rádio Ecclésia, said the state daily, 'with a view to mobilising the population to [commit] anti-social and anti-government acts'. On 10 July, Rádio Ecclésia journalist **Alexandre Cose** was barred from covering the forced movement of 13,000 families from Boavista to a tented camp some 40km from Luanda. Two days later BBC correspondent **Justin Pearce** and **Rafael Marques** (*Index* 2/2000, 3/2000, 5/2000, 6/2000, 2/2001), a local freelance journalist and human rights activist, were also barred from the camp. An official from the Angolan Ministry of Social Reinsertion said: 'We are aware that there is freedom of press in some countries, but we doubt if it applies to our situation here.' (MISA, OSF)

## ARGENTINA

On 6 June a court in Santiago del Estero in north-west Argentina brought a third seizure order against the local paper *El Liberal*. After 11 complaints by members of the Women's Branch of the Peronist Justicialista Party the seizure orders now total 600,000 pesos (US$600,000), the paper's future is in doubt. 'It is clear that this a campaign by the provincial government, which is using a justice system that is not independent to silence a press outlet in reprisal for its critical reporting,' said IAPA President **Danilo Arbilla**. *El Liberal* and the Cordoba daily *La Voz del Interior* published reports critical of the Women's Branch in July 2000, triggering the campaign against both papers. (IAPA, IFEX)

## ARMENIA

Journalist **Vagram Agadzhanyan**, forced to leave Stepankert in Nagorno-Karabakh after he was sentenced to a two-year conditional discharge for slander, found himself and his family under intense surveillance by the local militia in June. But when he moved to the capital Yerevan, the militiamen followed. To shake them off Agadzhanyan's sister arranged for the Yerevan newspaper *Aravot* to print the names and telephone numbers of some of the special agents who had been trailing her brother. (CJES)

## AZERBAIJAN

On 18 July, independent ABA TV president **Faik Zulfugarov** (*Index* 6/2000, 1/2001) announced that he is seeking political asylum in the USA. Breaking a two-month silence, Zulfugarov, 27, told 30 journalists during a 45-minute phone-in press conference that authorities have pressured his station, 'because we were independent'. Azeri authorities say Zulfugarov must face charges of slandering the country's president. Zulfugarov, once one of Azerbaijan's most powerful independent broadcasters, announced that his station was to close in June. Within hours the authorities seized the station's equipment in a 1am raid. (CJES, IFEX, RFE/RL)

The chief editor of the *Hurriyyat* newspaper in Baku, **Suleyman Mammadli**, (*Index* 3/2001) blamed financial difficulties and state pressure for its closure in early July after ten years in print. Also in July, the Azeri daily *Nedelya* reported that that four regional TV stations had been officially warned for 'unauthorised' broadcasting of television programmes. The TV stations say they repeatedly applied for licences without response. (CJES, RFE/RL)

On 19 July, journalist **Rahim Namazov** of the weekly *Eliller* was sentenced to six years in jail. He had been attacked and arrested by police who later destroyed his camera and tape recorder and charged him with taking part in an illegal 'Society of Karabag' invalids street rally and attending meetings. Namazov says he was only routinely reporting for his newspaper. (JuHI)

Azeri President Heidar Aliyev ordered all the newspapers in Azerbaijan to switch once and for all from Cyrillic script to the officially approved 32-letter Latinised Azeri alphabet.

The handover was 1 August. Nearly all Azeri newspapers used Cyrillic or a mix of the two before the deadline, as most people over 30 still prefer the old type. Many papers fear that they will now lose circulation to the broadcast media, a particular threat to cash-squeezed independent papers. (CJES, IWPR)

On 4 September the independent newspaper *Baku Boulevard* was closed by a court order, following a complaint by Baku Mayor Hajibala Abutalibov, who claims he was libelled in a 16 June article titled 'The State Racket'. The court also sentenced editor-in-chief **Elmar Huseynov** to pay 80 million manats (about US$17,400) in damages. The size of the fine breaches Azeri media law rules restricting damages payments to no more that three times its monthly turnover, to prevent the use of excess fines to close critical publications. (CJES)

## BANGLADESH

**Seik Selim**, correspondent for the daily *Bhorer Kagoj*, was threatened by fundamentalist Jamar-e-Islami leader Abdul Ali from Samanta bazar in the country's south-west, following a 6 May report in which he criticised Ali's group for demolishing parts of a monument to the country's language movement. **Qamruzzaman Sohel**, Faridpur district correspondent for the daily *Manavjamin*, was threatened in mid-May after he wrote a number of stories about criminal activity in the region. Following a 12 May story in which 14 members of the group were named and attempts were made to find out who the leader was, Sohel was told that he would have his legs cut off if he continued with his investigations. (*Media Watch*)

## BELARUS

The run-up to the 9 September presidential elections, saw a full-scale state-sponsored assault on independent voices in the country in a bid to guarantee re-election for the incumbent, Alexandr Lukashenko.

On 12 July, police seized equipment, some of which was leased under a US government aid programme, from the Krichev-based weekly *Volny Horad* for violating Decree No. 8, which bars the use of foreign grants for activities that encourage 'agitation'. On 20 July, according to local press reports, officials from the Markovka village prosecutor's office, accompanied by police, seized equipment from the weekly *Belaruski Ushod* provided by the international NGO IREX, which later won a court order on 15 August to have the equipment returned five days later. On 17 August, police seized 400,000 copies of the independent tri-weekly *Nasha Svaboda*'s pre-election issue, which endorsed **Vladimir Goncharik**, the only opposition candidate. On 21 August officials seized computers and other equipment from the independent daily *Narodnaya Volya*. On 27 August, authorities moved swiftly to install a State Press Committee official as acting director at Magic, the last remaining non-state printing press (*Index* 6/2000, 1/2001). The effect was immediate: on 28 August, 40,000 copies of a special issue of *Rabochi* were seized and the editor, **Viktor Ivashkevich**, was charged with slandering the president. On 31 August, Magic printed a special edition of the independent newspaper *Predprinimatelskaya Gazeta* with two blank sections. On 5 September, *Narodnaya Volya* was allow to print only after a collage titled 'Lukashenko is the past, Goncharik is the future' was removed from its front page. Lukashenko recorded a disputed landslide victory on the day; yet ironically the official media barely reported it, preferring instead to discuss a construction expo and criticism of US policy in Israel, apparently in a bid to gloss over the weeks of wrangling. International observers concluded that that vote 'failed to meet international standards'. (CPJ, IFEX, RSF, Article 19, BBC)

## BOTSWANA

On 9 July the warring sides in the long-running dispute between the Botswanan government and the *Botswana Guardian* and *Midweek Sun* finally got their first day in court. The government has withdrawn state-funded advertising from the two papers in punishment for their critical reportage. However, the two papers say the order had been extended to a ban on purchase of the papers by government officials and departments. As a side effect, they said, government officials were refusing direct interviews, invitations to press reception were being withdrawn and routine services like police reports were being withheld. In an opening hearing the two papers convinced Justice Isaac Lesetedi to issue an interim order declaring that the April advertising ban did not extend to purchasing the papers and did not include a news blackout. The hearings into the advertising ban itself continued. (MISA, IFEX)

## BRAZIL

On 21 July magistrate Ana Paula Braga Alencastro ordered the seizure of the 22 July edition of the daily *Tribuna Popular* in the southern city of São Lourenco do Sul. It had led on a report that city mayor Dari Pagel had been charged with five others with fraudulent management of the city's pension and healthcare funds. The magistrate said the paper 'could not manipulate public opinion and denigrate the image of a public figure'. Three other papers ran the story anyway, triggering huge sales among city employees. (RSF)

## BRITAIN

An injunction shielding the identities of the recently released killers of James Bulger was altered on 10 July to protect Internet service providers (ISPs) from prosecution if banned material is posted on their servers without their knowledge. (*FT*, *Guardian*)

On 11 July the government published the Office of Communications bill, aimed at replacing the six present telecommunications and broadcast regulators with a new single regulatory body, *Ofcom*. BBC director-general **Greg Dyke** warned of the danger of relying on a single body to regulate the content of so broad a range of media. (*FT, Guardian*)

Three government ministers attacked Channel 4 for airing a spoof documentary about media response to paedophilia, *Brass Eye*, on 25 July. Secretary of State for Culture Tessa Jowell also challenged regulators at the Independent Television Commission's failure to act before Channel 4 repeated the programme two days later, and suggested that the TV regulatory body should be given stronger powers to prevent repeats of programmes which are the subject of large numbers of public complaints. Channel 4 withdrew the programme from its digital and satellite station, E4. A government spokesman insisted that it did not see itself as a 'state censor' of television. (*Guardian, Daily Telegraph, The Times*)

A High Court judge criticised London Underground (LU) for imposing a 'gagging' order on London transport commissioner **Bob Kiley**. Justice Sullivan ruled on 31 July that Kiley could publish an edited version of an independent report allegedly critical of the government's Public Private Partnership (PPP) financing scheme for the London tube system. LU had argued that the report contained commercially sensitive material and that publication would breach a confidentiality agreement signed by Kiley. (*Guardian*)

On 31 July **Louis Farrakhan**, leader of the African-American group the Nation of Islam, succeeded in overturning a British Home Office exclusion order that has barred him from Britain for 15 years. The Home Office had argued that Farrakhan's history of anti-Semitic and racist remarks could increase inter-community tension. His lawyers successfully countered that the ban was in breach of the right to freedom of expression, guaranteed under the Human Rights Act. Farrakhan will continue to be excluded pending an appeal by the Home Secretary. (*Daily Telegraph, Guardian, The Times*)

Canadian mining giant Barrick Gold sued *Observer* columnist **Gregory Palast** and Guardian Newspapers Ltd over a November 2000 article on the corporate funding of US President George W Bush's electoral campaign, in which Palast cited an incident at a Barrick subsidiary in Tanzania in which several people were alleged to have been killed (the case had been investigated by Amnesty International, but AI has said it could not substantiate the claim). Barrick demanded the removal of the offending portion from Palast's US-based website as a condition of a UK court settlement in which it received an undisclosed 'substantial sum' from the newspaper group. The full article is still available on several websites unrelated to Palast or *The Observer*, including www.littlegeorgebush.com and www.therant.net (*Observer*)

In July Britain's Police Federation repeatedly sent legal warnings to cinemas planning to screen a documentary about deaths in police custody. The

Federation claimed that accusations of murder by victims' relatives in the film *Injustice* had no legal basis and could result in cinemas having to pay substantial libel damages. The screenings' organisers complained that the police's solicitors had in each case issued the warnings at the last minute, leaving no time for the cinemas to seek their own legal advice. (*Guardian, Observer*)

On 7 August the British Home Office abandoned a 'pre-entry clearance' scheme operating in the Czech Republic. Under the scheme, said to discriminate against Roma travellers, British immigration officials in Prague screened passengers travelling to London before departure. A Czech TV documentary filmed two undercover journalists, one of Roma origin, trying to board a flight; the Roma was prevented from travelling despite giving the same answers to the British officials as his colleague. The recently passed British Race Relations (Amendment) Act, allows immigration officers the right to discriminate against certain specified ethnic groups, based on the state's assumption that they are more likely to claim asylum on arrival in the UK. (*Guardian*)

In August prisoner **John Hirst**, who founded The Association of Prisoners from his Nottingham jail to campaign for prisoners' rights, including the right to vote, opened a case in the High Court challenging prison rules barring inmates from speaking to the media. After Hirst was heard giving a radio interview he was moved to a new prison where his calls could be controlled. He argues that the rules are in breach of the right to freedom of expression enshrined in the Human Rights Act. (*Daily Telegraph*)

After the August publication of *Bitch* by children's author **Melvin Burgess**, a novel for teenagers about a sexually active 17-year-old who turns into a promiscuous dog, the Library Association and publishing trade magazine *The Bookseller* announced that they had received calls from parents and teachers for a national ratings system for children's fiction. (*Daily Telegraph, Observer*)

On 15 August the Court of Appeal allowed the anti-abortion group **Pro-Life Alliance** to challenge a BBC ban on a party political broadcast showing images of aborted foetuses. The High Court upheld the BBC's ban in June on the grounds that the images offended public taste and decency. The group had challenged the decision under the right to freedom of expression under the European Convention on Human Rights. (*Guardian*)

## CANADA

The Royal Canadian Mounted Police confiscated and copied tapes made by Aboriginal People's Television Network reporter **Todd Lamirande** after he filmed a 24 June clash between supporters and opponents of a planned ski resort in Sun Peaks, British Columbia. The network is suing for illegal detention, search and seizure. (IFEX)

An exiled Pakistani journalist living in Toronto faced death threats after his articles in a community paper raised the ire of some Muslims in the city. **Tahir Aslam Gora**, editor and publisher of the local weekly *Watan*, had published an item about Muslim women's rights early this year that included allegations against Toronto Muslims in polygamous marriages. He received a death threat over the phone and warnings from people in his local mosque. The threats, including a bomb scare, continued into July after he published articles critical of the Afghan Taliban and reprinted a local newspaper story about a Muslim school principal charged with sexually assaulting students. (CIFE, IFEX)

## CAMEROOON

**Albert Mukong**, the Cameroonian journalist and human rights activist who took his case to the UN Human Rights Committee in 1994, was awarded US$137,000 compensation on 11 June in recognition of the abuses he suffered at the hands of the authorities. Mukong, a long-time opponent of the one-party system in Cameroon, was detained in 1988 and 1990. His book about his time in jail, *Prisoner Without a Crime*, was banned and he was forced to leave the country. (RSF, IFEX)

On 30 July police seized copies of the independent newspaper *Mutations* after it published a list of 21 leaked decrees on military reforms, signed by President Paul Biya on 26 July but not publicly released. Publisher **Ahman Mana** refused to name his sources, citing a December 1990 law that allows him that right. The same law failed to protect **Georges Baongla**, from the weekly *Le Démenti*, on 22 August. After accusing economic minister Michel Meva a M'Eboutou of embezzlement on 14 July, the police demanded he reveal his sources. Baongla refused, so they jailed him for allegedly failing to repay 500,000 CFA francs (about US$695) loaned him by a minister's nephew. Baongla denied the charge. (RSF, IFEX)

On 20 August, Radio Silantou journalist **Rémy Ngono** was assaulted by police officers after he criticised the police in his popular daily programme *Coup Franc*. There was a similar assault by police in May. (RSF, IFEX)

## CHILE

Chile's new press law, signed into force on 18 May, drew early criticism. While celebrating an end to old restrictions, IAPA president **Danilo Arbilla** said the new law introduced a new regulatory framework 'created to curtail the practice of journalism'. The law establishes a right to the protection of sources, but restricts this right to 'recognised' journalists, who must have a university degree in journalism. 'We are dismayed by the new law's attempt to create a legal distinction between journalists who may protect their sources, and journalists who may not,' said CPJ director Ann Cooper. (IAPA, CPJ)

The new law's effectiveness was soon tested in the case of **Alejandra Matus**, exiled in April 1999 after her book *El Libro Negro de la Justicia Chilena* [The Black Book of Chilean Justice] was banned and seized under Article 6b of the now defunct state security laws (*Index* 4/1999, 6/1999, 2/2001). The detention order was at first upheld, but dropped on appeal and she returned to Santiago on 15 July. Matus is now following a complicated trail through the Chilean Supreme Court to have the ban lifted. The book tackles widespread incompetence, bias and corruption in the Chilean judicial system. Justice Jordan, who brought the original defamation case against Matus, filed an appeal on 30 June against the partial dismissal of the case. Matus's lawyer, her brother Jean Pierre, then filed an appeal on 25 July urging the judge to dismiss the case entirely. The Supreme Court is expected to rule on both appeals within five months. (CPJ, IPYS)

On 10 July Chile's congress amended the constitution to eliminate film censorship, replacing it with a rating system that 'emphasises the right to unhindered artistic creation and expression'. The ratings council, however, still comprises representatives of the four armed services, the police and the legislature. Perhaps with them in mind, a scheduled 11 July private screening of a film by **Martin Scorsese**, *The Last Temptation of Christ*, banned in Chile since 1997, was cancelled at short notice for fear that the cinema would be closed and the audience arrested. (IPYS)

Fundación Terram, a Chilean NGO for sustainable development, won the first round in a legal bid to force the governmental National Forestry Corporation (CONAF) to produce private papers responding to its claim that timber companies were breaching forest management rules. Terram's critical analysis of the work of four companies in the south of Chile in 1998 was submitted to CONAF for action, but when nothing happened, Terram cited the country's Administrative Integrity Law to demand copies of letters proving CONAF had taken action. The motion culminated in a 12 June ruling that rejected CONAF's plea of privacy rights and ordered the agency to hand over the papers in ten days. CONAF is appealing against the decision. (IPYS/IFEX)

## CHINA

A student known only as **Liu** was given a four-year suspended sentence and fined 2,000 Rmb (US$248) by a court in Chengdu, Sichuan Province on 4 June, for uploading pornographic material on to the Internet from his home PC. Despite elaborate precautions, Liu is thought to be the first Internet user to have been traced to his home, rather than through an Internet café, reflecting new techniques and technology employed by police. Elsewhere on the Net, web newsletter *Hot Topic*, supplying 235,000 subscribers with news and comment four times a week, was banned on 18 June and an online chatroom on the *Xici Hutong* website titled *Democracy and Human Rights* was closed on 22 June.

The latter site was reported to have recorded a million hits a day. Following repeated diplomatic complaints, access to the Australian government's foreign affairs website was restored on 4 July after 18 months of blockage. Beijing claimed 'technical problems', though the site briefly came back online during a visit by Australia's communications minister in June. The site contains details of human rights developments and consular warnings to tourists travelling in certain areas of China. *Xinhua* reported on 19 July that 2,000 Internet cafés in China had been closed and another 6,000 ordered to 'make changes' after a three-month official investigation into their activities. Internet cafés are required by law to keep records of customers' browsing habits and bar access to government-proscribed sites, though commercial imperatives may lead many café owners to try to bend the rules. Human Rights in China (HRIC) released a report on 2 August condemning 60 new sets of rules on Internet content in China, under which 14 people have been jailed for expressing personal views online since January 2001. The report urged multinational computer firms sponsoring the 2008 Olympics to work towards lifting the controls. The trial of **Huang Qi** (*Index* 4/2000; 5/2000; 2/2001) was again postponed on 27 June 'because of the Chinese Communist Party's 80th birthday', said an official at the court in Chengdu, Sichuan Province. Huang faces charges of subversion for posting articles critical of the government on the Internet. Arrested in June 2000, Huang's original trial in February was suspended when he collapsed in the dock showing signs of physical abuse. It is widely thought the June trial was postponed to avoid compromising Beijing ahead of a meeting in Moscow on 13 July to decide the venue of the 2008 Olympics. His trial was finally completed in a closed session on 14 August, although the verdict and sentence are yet to be announced. Huang's wife, **Zeng Li**, tried to take her husband's photograph in a corridor of the courthouse, but security officials confiscated the film. (AFP, AP, CPJ, HRIC, Reuters, RSF, Xinhua)

**Shen Zhidao** (*Index* 4/2000), twice arrested on Tiananmen Square for staging one-man pro-democracy protests on 4 June 1999 and 2000, was tried for subversion in Shenyang, Liaoning Province, on 18 June. Sentence is yet to be passed pending reports on Shen's mental health. Businessman **Li Hongmin** was detained in the middle of June after he was found to have emailed friends several Chinese versions of *The Tiananmen Papers*, the inside story of the political lead-up to the 1989 massacre. He is presently in detention in his hometown of Shaoyang in Hunan Province and faces a possible ten-year jail sentence for subversion. (AFP)

**Roger Parkinson**, president of the World Association of Newspapers (WAN) criticised China for its treatment of journalists and excessive control of the media in his opening address to the association's conference in Hong Kong on 4 June. Hong Kong Chief Executive Tung Chee-hwa responded by claiming that journalists were free to travel in China and 'hear the voice of the people everywhere'. (WAN)

Reports emerged on 5 June that **Qian Gang**, editor-in-chief of the Guangzhou-based *Southern Weekend* and his front-page editor **Chang Ping** were dismissed by the State News and Publishing Bureau (SNPB). It is thought the paper's coverage of rural unrest, official corruption and crime had violated propaganda regulations. A forum on the paper's website was closed on 18 June after contributors criticised the government's decision to fire the editors.
**Ma Yunlong**, vice editor-in-chief of *Dahe Bao* in Zhengzhou, Henan Province, was removed from his post for approving stories exposing two separate cases of official corruption, said a spokeswoman for the paper on 17 June. The paper's editor-in-chief, **Ma Guoqiang**, was disciplined but kept his job.
The *Business Morning Daily* in Nanjing, Jiangsu Province, was closed on 22 June after accusing President Jiang Zemin of granting preferential policies to his hometown of Shanghai. Propaganda officials called the article a 'grave political error'.
Finally, reports emerged on 27 June of a document circulating among China's major newspapers and periodicals warning publications to toe the Party line or face instant closure. Criticism of the Party or calls for political reform is strictly forbidden, as is speculation about national leadership

changes. The objective is reportedly aimed at 'maintaining the correct opinion orientation and preventing the corrosive influence of a western outlook on journalism'.

**Jiang Weiping**, a reporter for the pro-Beijing Hong Kong newspaper *Wen Wei Pao* was reportedly jailed for four years on 28 June for 'leaking state secrets'. Jiang wrote two articles in 1999 for a Hong Kong magazine exposing corrupt dealings by the governor of Liaoning Province, Bo Xilai, and his family and friends.

Ms **He Qinglian**, a blacklisted journalist and scholar (*Index* 5/2000) announced on 30 June that she had sought exile in the US after facing intensifying harassment by police in Shenzhen (*Index* 3/2001). 'I was afraid they were gathering evidence with the aim of locking me up,' she said.

Editors from 180 periodicals were summonsed to a conference in Beijing on 6 July, and told to expect a 'shake-up' in the sector, perhaps involving a large-scale amalgamation of several titles under Party control. **Yao Xiaohong**, a journalist on the *Jiangxi Metropolitan Consumer News,* was fired in July after he wrote an article in April about the removal of an executed prisoner's organs for transplant. The piece was published on the *People's Daily* website, but after the US Congress drew attention to this so-called 'organ harvesting' and Beijing decided to deny the practice officially, Yao was found and fired for 'violating editorial rules'.

President Jiang Zemin was thought to be personally behind the closure of two Communist Party journals, *Zhenli de Zhaoqiu* (Seeking Truth) and *Zhongliu* (Mainstream), which criticised Jiang's recent decision to allow capitalists to join the Party, and accused him of building a cult of personality around himself. (AFP, AP, CPJ, *Hong Kong Mail*, ICHR, Reuters, *South China Morning Post*, WAN)

The Tibetan Centre for Human Rights and Democracy (TCHRD) reported on 16 June that a woman called **Migmar** and four friends had been arrested at home in May for watching a video of the Dalai Lama. The four friends were released several days later after paying US$620 fines, but Migmar was sentenced to six years in prison on charges of 'inciting separatism'.

Hundreds of Tibetans were arrested in the days running up to the Dalai Lama's birthday on 6 July in an apparent attempt by authorities to halt illegal celebrations. Six Tibetans were sentenced to prison terms ranging from seven years to life in separate trials, it was reported on 30 July. (AFP, AP, Reuters, TIN, TCHRD)

The Religious Affairs Bureau (RAB) in Sichuan Province forcibly evicted thousands of monks, nuns and students from the Serthar Tibetan Buddhist Centre in mid-June before destroying most of the centre's accommodation buildings. The RAB reportedly had wanted to reduce the centre's student population from 10,000 to 1,300 since June 1999. Students were ordered to leave, said an RAB spokesman on 21 June, because of 'concerns about social stability and the order of central authorities'. **Khenpo Jigme Phuntsok**, 68, the monastery's abbot, is thought to be in detention. (AFP, AP, Reuters, TIN)

**Steven Shaver**, a US photographer working for AFP, was assaulted by six police officers on 23 June outside the Forbidden City in Beijing as the 'Three Tenors' gave a concert to promote the city's Olympic bid. Shaver was photographing the arrest of a lone political protestor when he was attacked. A senior officer called them off, but the same group later attacked Shaver again as he left the concert. The US State Department, AFP and RSF lodged strong complaints with Beijing, and with senior Olympic officials. Foreign ministry spokeswoman Zhang Qiyue said that though Shaver had 'provoked' the officers by photographing the arrest of what she claimed was a 'ticket tout', the 'physical contact' was inappropriate and the officers 'did not keep their cool at certain points'. (AFP, AP, Reuters, RSF)

## COLOMBIA

On June 27 **Pablo Emilio Parra Castanada**, founder and head of *Planadas Cultural Estereo* community radio in the village of Planadas, Tolima province, was found dead in an area known for opium poppy plantations controlled by the Revolutionary Armed Forces of Colombia (FARC). Parra was the president of the local Red Cross and his radio programmes included broadcast school lessons for those too isolated or too poor to attend school. The Colombian army

claim he was assassinated by FARC, and it is not clear whether Parra's death was related to his work as a journalist.
Also in Tolima province, in Fresno, the director of *Fresno Estereo* local radio, **Arquimedes Arias Henao**, was shot dead in the station building on 4 July. Arias mostly covered community and cultural affairs and the motive for his murder remains obscure.
On 6 July, **Jose Dubiel Vasquez**, director of *La Voz de la Selva* radio in the city of Florencia, in Caqueta province, was shot dead by two men on a motorcycle on his way home. There have been many clashes in the province, but Vasquez had not received any threats up to his death.
**Jorge Enrique Urbano Sanchez** was shot four times by two men in a bar in the city of Buenaventura on 8 July. Urbano was a radio reporter for Emisara Mar Estereo, a local radio station, and had earlier received death threats from an unidentified group.
**Eduardo Estrada Gutiérrez** was shot dead on 16 July after attending a family gathering in San Pablo, in southern Bolívar, eastern Colombia. He had been working on a new community radio station project. The authorities have no suspects, but IPYS cited local opinion that the assassins are right-wing paramilitaries from the United Self Defence Forces of Colombia. (IAPA, IFJ, CPJ, RSF, FLIP)

## COSTA RICA

On 7 July radio journalist **Parmenio Medina** was shot on his way home from recording his radio programme, *La Patada* (The Kick) a political satire show on Radio Monumental in the capital San José. The Costa Rican government condemned the murder and launched an investigation. In April, *La Patada's* broadcasts were suspended and the station closed after complaints from 'Church authorities' about Medina's criticism of the management at Catholic radio station *Maria de Guadalupe*. (IAPA, IFJ, RSF)

## CONGO (DRC)

**Freddy Loseke Limsumbu La Yayenga**, publisher of *La Libre Afrique*, was accused of defaming evangelist pastor Sonny Kafutu of the Army of the Eternal Church in a report alleging non-payment of debts. 'Accuse me of anything,' the pastor reportedly said, 'I am the brother of the person in power. Nobody will do anything to me.' Loseke (*Index* 2/2000, 3/2000, 4/2000) was released from a year in jail in January.
Also reported in June: the detention of **Joachim Diana Gikupa**, publication director of the daily paper *L'Avenir*, over an article headlined 'Kabila torpedoed by his entourage?' which cited a fax from President Laurent Kabila's cabinet director, later claimed to be a forgery. (JED)

## COTE D'IVOIRE

On 20 June, two unidentified gunmen entered the home of **Laurent Tapé Koulou**, editor of the independent daily *Le National*, attacking his family with an automatic weapon when they failed to find him. The journalist's older sister, **Tapé Ziadou Madeleine**, and a friend, **Takoré Clovis Désiré**, were killed and a third person, **Tapé Serge Médard**, seriously wounded. At the time of the attack, Tapé Koulou was in France to cover a visit by Ivoirian President Laurent Gbagbo. (RSF)

**Abdoulaye Sangaré** (*Index* 2/2001), managing editor of the independent daily *Le Jour*, was informed by sources on 21 July that a secret 50 million CFA (US$69,400) bounty had been put on his head. He is considered to have 'sensitive information' about his president. **Patrice Guéhi**, publisher of *Le Patriote* and supporter of the opposition Rally for Republicans (RDR) of Alassane Ouattara, also reported getting death threats from the country's notorious Territory Supervision Office. (RSF)

## CUBA

No change in the status of labour activist and Internet journalist **José Orlando González Bridón**, who is still held in Combinado del Este prison, east of Havana. His charges, however, do change. He was initially charged with 'spreading false information' in articles critical of police negligence in the death from domestic abuse of his trade union colleague **Joanna González Herrera** on the *Cuba Free Press* website. In April 2001 his charge was changed to bring it under Article 103 of the Cuban Penal Code, which threatened to prolong his sentence for up to 15 years. On 24 May Bridón's lawyer convinced the court of the accuracy of

Bridón's story and the fact that the website could not be seen by the average Cuban, for whom Internet access is government-controlled. Accordingly the prosecutor changed the charge to 'defamation' under Article 204 of the code and demanded a one-year sentence. On 8 June, the charge changed again, to Article 115 of the code, and he got two years in jail for spreading 'false information for the purpose of disturbing the international peace, or to endanger the prestige or credibility of the Cuban State or its good relations with another State'. (RSF)

On 5 August independent journalist **Jadir Hernández** was put under house arrest in the town of Guines after a series of threats over his work as a correspondent for the independent Havana Press agency. Also in late July and early August three independent journalists belonging to the Manuel Márquez Sterling Journalists Association – **Jorge Olivera Castillo**, **Graciela Alfonso** and **Jesús Alvarez Castillo** – were beaten and interrogated by security agents in a similar bid to stop them from reporting. (IAPA)

## CZECH REPUBLIC

On 15 June, Czech journalist **Tomas Smrcek** was found not guilty in a case brought under the Czech Secrets Act, held behind closed doors since mid-November 2000. Smrcek was on trial for displaying confidential government documents during a 1994 television broadcast that proved how a political candidate had improperly tried toclear a friend of drunk-driving charges. (IFJ)

## EGYPT

On 14 June Egypt's journalists' syndicate announced an investigation into three members – **Gehud Auda** and **Hassan Fouad** from the state-run *al-Ahram*, and **Ashraf Radi** of *al-Alam al-Yom* – who broke syndicate rules on contact with Israelis by attending a reception at the Israeli embassy in Cairo. (*Cairo Times*)

**Mamduh Mahran**, chief editor of the independent dailies *al-Nabaa* and *Akhir Khabar*, was detained for two days on 17 June and charged with offences including religious sedition and printing indecent images following his papers' front-page articles that day about a Coptic Christian monk alleged to have had sexual relations with women then extorting them. If convicted, Mahran could face up to five years in prison. The report drove thousands of Christians to protest outside the former monk's monastery and the Abbasiyya Cathedral in Cairo, leading to clashes with the police. (IPI, *Cairo Times*)

An administrative court finally declared the Egyptian Organisation for Human Rights (EOHR) legal on 1 July. Created in 1986, it has operated more or less illegally ever since, despite repeated attempts to register itself as a NGO. In 1997 it was refused a licence on the grounds that other organisations dealing with human rights already existed. (*Cairo Times*)

The Supreme Security Court in Cairo released a ruling on 19 June jailing Professor **Saad Eddin Ibrahim** and 27 other members of the Ibn Khaldun Research Centre. Ibrahim received seven years in prison with hard labour; two others got five years, and four two years' hard labour. All the others received one-year suspended sentences. Prof. Ibrahim, a sociologist, and the other defendants had undertaken and published a study that uncovered irregularities in the 1995 parliamentary elections. The research was funded by a grant from the European Union. (*Cairo Times*)

Legal proceedings to dissolve the marriage of writer **Nawal al-Sa'dawi** (*Index* 9/1987, 9/1990, 8/1992, 3/2001), were dismissed on 29 July. The case was instigated by Nabih al-Wahsh under *hisba* (society) law, a rarely applied Islamic concept which allows any Muslim to charge another with apostasy, as a result of comments Sa'dawi allegedly made about the practice of Islam. (*Guardian, Cairo Times*)

The Supreme Mufti office in Cairo, Egypt's highest religious authority, denounced the Egyptian version of the TV game show *Who Wants to be a Millionaire?* as a sinful form of gambling and the work of Satan, the BBC reported from Cairo on 3 July. Though it is a quiz and not a game of chance, the Mufti argued that it encourages gambling because competitors have to spend money on phone calls to answer questions without knowing if they will win the chance to go on the show. (BBC)

## ERITREA

There was continued concern during the summer for the safety of 15 Eritrean journalists allegedly jailed or forcibly conscripted, among them **Paolos Zaid**, assistant editor of the government weekly *Eritrean Profile*, picked up in April; **Zemenfes Haile**, a founding editor of the independent weekly *Tsigenay*, arrested and tortured last year; **Ghebrehiwet Keleta**, a freelance working for *Tsigenay*, **Temesghen Ghebreyesus**, a reporter for the independent *Keste Debena*, who is believed to have been drafted and sent to the Eritrean-Ethiopian border. (CPJ)

## ETHIOPIA

**Tamrat Zuma**, former publisher and editor-in-chief of the closed Amharic weekly *Atkurot*, arrested on 25 May, remains in jail charged with incitement to rebellion over a two-year-old article. The item had quoted a retired Ethiopian general quoted in turn by the US-based magazine *Ethiopian Review* warning of the 'imminent' overthrow of the current government.

**Lubaba Said**, editor-in-chief of the independent Amharic weekly *Tarik* in August was also charged over a two-year-old item that alleged that government security men had abandoned their posts. Said was released on 20 July after posting bail and her trial adjourned. Earlier that month police had called in seven editors of independent Amharic newspapers for questioning. They included: **Berhan Hailu** of *Wogahta*, **Merid Zelleke** of *Satanaw*, **Mengistu Wolde Selassie** of *Moged*, **Leyikun Ingida** of *Dagim Wonchif*, **Henock Alemayhu** of *Medina*, **Daniel Abraha** of *Netsanet* and **Tilahun Bekele** of *Netsabrak*. The papers had reported claims that Foreign Minister Seyoum Mesfin was planning to abandon the ruling Tigray People's Liberation Front (TPLF) to join a splinter faction. Ingida, Alemayhu, Abraha and Bekele were released on bails ranging between 3,000 and 5,000 birr (US$400–650). Hailu was detained for 48 hours after her interrogation. Zelleke and Selassie remained in custody until 13 July, while police pressed them to reveal the sources for their articles. Prosecutors are currently preparing to file charges against all seven journalists, according to local sources. (CPJ, WiPC, PEN)

**Solomon Nemera**, editor-in-chief of the former magazine *Urjii,* was released in June, with all charges dropped. He had spent a year in jail and two more years before that awaiting trial after his arrest in October 1997, after *Urjii* claimed that government troops killed three alleged members of the Oromo Liberation Front. (CPJ, WiPC, PEN)

## FRANCE

On 2 June, the offices of the weekly *Le Canard Enchaîné* office in Paris was broken into. Two cabinets were searched and the laptop computer of journalist **Brigitte Rossigneux**, who deals with defence issues, was stolen. The journalist believes the raiders were after clues to her contacts' names, especially those at the Ministry of Defence. (RSF)

## GEORGIA

Popular 26-year old Georgian journalist **Georgy Sanaya** was found dead in his Tbilisi apartment on 26 July, shot in the head at close range. Sanaya anchored *Night Courier*, a nightly political talk show on the independent television station Rustavi-2. **Erosi Kitsmarishvili**, executive director of Rustavi-2, told the CPJ that the station has been a frequent target of government harassment in recent years. Others suggested the involvement of gangs from Georgia's Pankiss Gorge area, scene of kidnappings and conflict between Chechen and Georgian groups. (CPJ, IFEX)

## GERMANY

In August Interior Minister Otto Schily called for the closure of websites containing illegal content under German law, specifically neo-Nazi and racist sites, mostly hosted by web companies in the USA. Schily planned meetings with US officials with a view to discussing the regulation, but it is believed the US's First Amendment right to free speech will override the civil laws he hopes to apply. (BBC)

## GHANA

Ghana's parliament voted unanimously on 27 July to repeal the country's criminal libel law, which had been used in the past to prosecute media professionals. The abolition of the law fulfils a campaign promise by President John Kufuor during December elections. But the first deputy speaker of parliament, Freddy

Blay, said that the absence of the law was not a licence for journalists to publish falsehoods. The government had announced in January that it would not use the law pending its deletion from Ghana's criminal code. (IRIN, Article 19, IFEX)

## GUATEMALA

On 8 June a Guatemalan court found four men guilty of the murder of **Bishop Juan Gerardi Conedera**, former director of the Archbishop's Office of Human Rights, killed on 26 April 1998 in Guatemala City (*Index* 3/1998). The judgement was a milestone in efforts to bring military and state officials to justice in a climate of 36 years of repression and intimidation. But within the month **Leopoldo Zeissig**, the Guatemalan chief public prosecutor who secured the conviction had been forced by repeated death threats to flee the country. (RSF)

## HAITI

The 9 June broadcast of his weekly political programme *Moment of Truth* and its 11 June rebroadcast triggered a string of threatening phone calls to Signal FM Radio journalist **Roosevelt Benjamin**. He had linked the newly founded Civil Society Majority Movement (MSCM) with families of senators from the ruling party, Fanmi Lavalas. Among a number of threats to the Haitian media, earlier this year armed men circled the neighbourhood of another Signal FM journalist, **Michel Soukar**, and questioned local residents. (CPJ)

On 20 June **Fritson Orius** of Radio Haïti Inter was assaulted by two armed men as he left the station. RSF, among others, link continuing pressure on the station's staff to last year's assassination of **Jean Dominique**, then the country's most respected political broadcaster. The fallout from the investigations into Dominique's death, especially since the indictment of ruling party senator Dany Toussaint in May 2001, continue to rock the Haitian media. Toussaint claims a plot and has called for the examining judge's arrest. During June and July his lawyers launched six actions before four separate courts in a bid to discredit or dissolve the inquiry body investigating the case. (CPJ, RSF)

On 9 August **Liberus Renald** and **Claude François**, journalists with Rotation FM in the north-eastern town of Belladères, were arrested on in the station's offices. Police demanded a tape of interviews with ex-soldiers accused of attacking police stations on 28 July. When they refused, the police beat them up and detained them for three hours. And after broadcasting criticism of the local authorities in the south-western town of Thiotte on 17 July, **Confident Fedner**, a reporter for the Catholic radio station Radio Sacré-Coeur, was repeatedly threatened by one of the mayor's security guards and later by a civil organisation close to the mayor and the ruling Fanmi Lavalas. (CPJ, RSF)

## HUNGARY

The 12 July appointment of **Karoly Mendreczky** as president of Hungarian Television (MTV) by its trustees upset some observers. Mendreczky resigned as a member of the ruling Fidesz party and president of its cultural committee to take the post, but this did not ease fears that the government was simply putting its own man in charge. The situation is complicated by the current make-up of the 'curatorium' set up under Hungarian broadcasting law to maintain the state broadcaster's independence. It is supposed to be a bipartisan body, but it presently only features government figures.

## INDIA

On 2 August novelist **Arundhati Roy**, lawyer **Prashant Bhusan** and environmental activist **Medha Patkar** made a defiant appearance before the Supreme Court where they face contempt charges relating to protests about the Narmada Dam (*Index* 2/1999, 1/2000, 3/2001). Roy presented an affidavit in which she protested about 'a disquieting inclination on the part of the court to silence criticism and muzzle dissent, to harass and intimidate those who disagree with it'. The Supreme Court is now considering new contempt charges against Roy. (IHT, AFP)

## INDONESIA

On 24 May journalists **Agus Wijanarko**, **Yon Daryon**, **Thomas Marsis**, **Bambang Mudjono** and **Sarjono** were attacked in Tegal while covering a rally held by pro-president Wahid Muslim organisation Laskar Diponegorol. The journalists' equipment was also destroyed. This is the 12th reported attack on journalists since May 2000 by supporters of the president. (AIJ)

*Serambi Indonesia*, the largest newspaper in Aceh Province, suspended publication after receiving threats from a separatist group to burn down the building and kill its staff. The newspaper feels unable to guarantee the safety of its employees and at present gives no indication of when publication will resume. (*Freedom Forum*)

## IRAN

**Morteza Firouzi**, former editor-in-chief of *Iran News*, and **Hamid Naïni**, of *Peyam Emrouz*, were released in July, but despite being pardoned in February **Mashallah Shamsolvaezin**, editor of the now banned Iranian daily *Neshat* (*Index* 1/1999, 1/2000, 6/2000), has not been released. Another *Neshat* journalist, **Latif Safari** (*Index* 6/1999) of *Neshat* was acquitted in June, but is also still behind bars.
Among other journalists detained in Iran, of the jailed reporters of *Iran-é-Farda*, relatives of **Reza Alijani** (*Index* 3/2001) have not been allowed to visit him since February, while there are fears for the health of Alijani's colleague in jail, diabetic **Hassan Youssefi Echkevari**. Another *Iran-é-Farda* journalist, **Ezatollah Sahabi** (*Index* 5/2000, 2/2001), who is over 70, is said by friends to have been submitted to psychological torture. (RSF)

On 16 July **Akbar Ganji** (*Index* 4/2000, 1/2001, 2/2001, 3/2001) was sentenced to six years in jail on charges of 'collecting confidential information that harms national security and spreading propaganda against the Islamic system'. Ganji's investigations into the murder of Iranian intellectuals and dissidents in 1998 implicated several top government officials. The 22 April charges stemmed from Ganji's participation in a controversial April 2000 conference in Berlin on the future of the Iranian reform movement. Another participant, **Khalil Rostamkhani**, a journalist with the *Daily News* and *Iran Echo*, was sentenced on appeal to eight years' imprisonment on 25 August. He had been sentenced in January to nine years for his part in the conference. (RSF, BBC, CPJ)

In the week of 30 July, the daily *Fath* was banned permanently after a previous six-month ban in April 2000. On 4 August, the reformist weekly *Farday-é-Rochan* had its licence withdrawn for 'publishing untruthful and libellous articles in breach of public decency'. **Davood Bayat**, the newspaper's editor, received a 4.5 million rial fine (US$2,600). On 8 August, the reformist weekly *Hambasteghi* was banned as a 'preventive measure' for publishing 'lies and insults'. **Gholamheidar Ebrahimbay Salami**, the newspaper's director, has had at least 16 complaints filed against him. (RSF, BBC, CPJ)

Scuffles broke out in July at a performance by Iranian comedian **Hamid-Reza Mahi-Sefat**, dubbed 'Iran's Mr Bean', in the city of Mashhad. Some 50 members of the Islamist *Ansar-e Hizbullah* group staged a religious protest at the show, but police 'drove the attackers out,' reported the official Iranian IRNA news agency, and the show finally started after an hour's delay. The group's hardline leader **Hamid Ostad** was arrested. (IRNA)

## IRAQ

The Shi'a leader and poet **Ayatullah Hussain Bahr al-Uloom** died in mysterious circumstances on 22 June. The Ayatollah had earlier refused to appear on the *al-Shabab* youth television channel, owned by Saddam Hussein's son, Uday, to congratulate Saddam's second son, Qusay, on two election triumphs. The Ba'ath government had executed, expelled or imprisoned all other clerical members of the family. (*Guardian*)

## ISRAEL

Senior officers suspended the Israeli army newspaper *BeMahaneh* for several weeks for articles published in the newspaper on 4 May that 'did not correspond to the army's norms'. An article published in the issue had profiled a homosexual reserve colonel. (RSF)

## ITALY

During the G-8 summit at Genoa between 20 and 22 July, police officers attacked journalists covering anti-globalisation protests. At around midnight on 21 July, the police raided the Genoa Social Forum, an umbrella group of anti-globalisation organisations including the Independent Media Centre (IMC). The police ransacked the IMC and searched the premises for film and photographs. The many injured among the media included **Sam Cole**, a Rome-based producer for AP TV; photographer **Timothy Fadek** of the Gamma agency; **Mark Covell** of Indymedia; photographers **Guy Smallman** and **Paul Matteson** working for Christian Aid; **Michael Gieser**, a Belgian journalist, who was beaten as he lay on the ground, and **Philippe Stein**, IMC member and journalist from Berlin, who was struck when he implored officers to stop the violence. Prosecutors have ordered media outlets to turn over photographs and audio/video tapes of the Genoa street violence. Italian law allows such orders without right to appeal, with stiff penalties on journalists who do not comply. As the CPJ noted, by forcing journalists to act as police informants, the law severely jeopardises journalists' safety and credibility. (CPJ)

## KENYA

Lawyer **Gibson Kamau Kuria** alleged in June that a fraudulent search warrant was used to search the premises of the country's troubled Citizen Radio and Television (*Index* 2/2000, 3/2000, 3/2001) The lawyer criticised state searches of private property that transcended constitutional limitations. (NDIMA)

## KOSOVO

Kosovo's Council for the Defence of Human Rights and Freedoms in Priština reported that the international community's appointed municipal administration in the Kosovan town of Mitrovica banned a play at the town's Sandër Prosi Theatre on 7 June, for failing to deliver a English translation of the script to the authorities before the show, so they could 'verify' its content. Instead the actors moved out of the theatre and performed it in the open air.

## KYRGYZSTAN

After initially approving applications for official registration for 16 new media registrations, the Kyrgyz Justice Ministry rescinded the certificates on 20 June, citing a hitherto unknown ruling requiring existing publications to re-register first. The effect was to block the publication of the new publications, which included new papers put together by independent publishers whose earlier papers had been forced to close by court action brought by the state. The matter was returned to the courts for a series of inconclusive hearings; meanwhile the state extended the re-registration deadline to 1 October, further delaying the new publications. The most prominent media outlets formed since 5 April includes two potentially influential newspapers: *Moya Stolitsa* and *Agym*. *Moya Stolitsa*'s editor is Alexander Kim, former editor of *Vecherny Bishkek*. *Agym* is a Kyrgyz-language newspaper seen as the direct successor to *Asaba* that was closed down in April 2001 because of lawsuits against the owner, **Melis Eshimkanov** (*Index* 5/2000, 1/2001, 3/2001). (*Eurasianet)*

On 18 July, a presidential decree revised Article 297 of the Kyrgyz criminal code to allow up to three years in jail and confiscation of property for 'persons who produce or distribute information intended to overthrow or undermine the constitutional order of the Kyrgyz Republic'. (RFE/RL)

**Samagan Orozaliev**, a local television journalist from Jalal-Abad, who was arrested on 28 May and charged with extortion, was transferred to the regional hospital on 31 July with severe heart problems. Orozaliev was arrested when police found US$300 on him that he had allegedly received after blackmailing local politician and businessman, Ergesh Torobaev. Orozaliev had been preparing material critical of Torobaev for publication on state television channel Zamana. Two days prior to his arrest he had filed a complaint with the police because of harassment by members of Torobaev's family. They had demanded he hand over the taped material and not make it public. Orozaliev believes that Torobaev's son slipped the US$300 into his pocket during a meeting with Torobaev on 28 May because five minutes later police arrived to arrest him on charges of extortion. (JuHI)

## LAOS

On 8 June Information and Cultural Minister Phandouangchit Vongsa said that new guidelines were being drafted in order to define which truth should be reported by the media and that penalties were being tightened for those who gave false information. Currently journalists deemed to write critical reports on the ruling party face jail sentences of between five and 15 years. (IPI)

## LEBANON

On 1 June a Lebanese military court opened a case against Lebanese-American citizen **Raghida Dergham**, New York bureau chief for the London-based daily *al-Hayat* and a noted commentator on Arab affairs. She did not appear at the hearing and a warrant has been put out for her arrest. The charges stem from her participation in a 19 May 2000 panel discussion at the Washington Institute for Near Eastern Policy on Middle East politics. The panel included Uri Lubrani, the former Israeli chief in then occupied southern Lebanon. The indictment accused Dergham of being 'a participant as a journalist in a debate that was arranged by a member of the enemy'. The court adjourned until 30 November to give Dergham the opportunity to appear. Observers believe the case may be driven by Dergham's critical coverage of Lebanon's handling of its dispute with the UN over the demarcation of the Lebanese-Israeli border. (AFP, CPJ)

On 9 August, two journalists were assaulted and a third detained in front of the law courts in Beirut as they covered a demonstration against earlier arrests of supporters of the exiled general **Michel Aoun** and other groups opposed to Syrian influence in Lebanon. AP photographer **Hussein el Moulla** and **Sami Ayad** from the daily *an-Nahar* were roughed up by Information Services agents when unidentified persons demanded that he hand over his film. He refused and was beaten by them until he escaped. **Yehia Houjairi**, a cameraman from the official Kuwaiti television station, was arrested by police officers as he filmed the demonstration. The president of the Photographers' Union had to intervene in order to secure his release a short time later. Later that day Minister of Information Ghazi Aridi said measures were needed to stop 'mistakes by media outlets which threaten state security'. (RSF)

On 16 August, Lebanese security officials arrested **Antoine Bassil**, a freelance working for the Middle East Broadcasting Centre's London-based radio service. Bassil is accused of serving as a go-between for Israeli officials and a Christian opposition figure named Tawfiq al-Hindi, arrested earlier that month. On 15 August Bassil and al-Hindi were charged with having illegal contact with Israeli officials. (CPJ)

On the evening of 18 August, Lebanese military intelligence agents detained **Habib Younes**, an editor for the London-based daily *al-Hayat*'s Beirut office, at his home in the town of Jbeil. The agents reportedly gave no reason for the arrest, and only told the journalist they wanted to ask him 'some questions'. It is believed that Younes is accused of either meeting with or planning to meet with an adviser to Israel's former coordinator for south Lebanon. (CPJ)

## LIBERIA

**Sam Dean**, editor of the *Monrovia Guardian*, was arrested and charged on 20 August with 'criminal malevolence' for publishing an article that claimed that police chief Paul Mulbah had been summoned for questioning by the country's parliament. Mulbah complained to the Press Union of Liberia (PUL) about the newspaper's 'sensationalism' and 'misleading' reports. (RSF)

On 2 July, Minister of Post and Telecommunications Emma Wuor informed Radio Veritas that it was no longer allowed to broadcast on shortwave radio, leaving Kiss FM and Radio Liberia International – both owned by President Charles Taylor – as the only stations that can air political news countrywide. Information Minister Reginald Goodridge said that by airing political programming, Radio Veritashad violated its permit, which only allowed the station to broadcast religious shows. Radio Veritas aired several shows critical of the Liberian government, including the often controversial programme *Topical Issues.* (CPJ, IFEX, RSF)

## MALAWI

On 28 June, Ombudsman **Enock Chibwana** awarded *Daily Times* acting chief reporter **Mabvuto Banda** and the late *Malawi News* editor **Horace Somanje** 30,000 kwacha (approx. US$395) for abuses that followed their arrested on 21 June 1999, following the publication of a *Malawi News* article that quoted opposition supporters encouraging the army to take over the government. The two journalists spent two days in jail, only to have the charges of inciting mutiny dropped a year later. In his ruling, the ombudsman stated that 'it is not a crime to report on events; it is absurd, backwardness and bad governance to arrest such neutral journalists'. The award came too late for Somanje, who died on 23 June. (MISA, IFEX)

On 26 July about ten suspected opposition supporters beat up journalist **Ken Ndanga**, of the controversial pro-government weekly tabloid *The Sun*, a bitter critic of the newly formed opposition National Democratic Alliance (NDA) backing sacked former senior minister Brown Mpinganjira. On the other side, members of the ruling United Democratic Front's (UDF) youth league were accused of assaulting **Brian Ligomeka**, correspondent for the South African agency African Eye News Service on 12 August. Ligomeka suffered a bruised jaw and leg. Two days later unknown gangs attacked newspaper vendors in the streets of Blantyre and took their copies of the independent weekly *The People's Eye* before invading the magazine's offices. Editor **Chinyeke Tembo** had to temporarily close the office.

## MALAYSIA

The Malaysia Chinese Association (MCA), the second largest party in the ruling National Front coalition, has acquired two major Chinese-language independent newspapers, *Nanyang Siang Pau* and *China Press*. Several editors have been asked to leave in an apparent bid to exert editorial influence. NF already owns virtually all the main Malay and English papers and the acquisition of these last two will further consolidate state control over the media. (CPJ)

## MOROCCO

**Alain Chabod**, deputy chief editor of the public TV channel France 3, was harassed by security forces as he tried to investigate new evidence regarding the 1965 disappearance of dissident leader **Mehdi Ben Barka**. Moroccan Secret Service (DST) agents began following him on 6 June. Later DST officers demanded a halt to the printing of the weekly magazine *Demain*, which contained some of the new evidence. Later that day, *Demain*'s editor, **Ali Lmrabet** (*Index* 1/2001), was photographed and verbally threatened, while with Chabod, by men identified by the Frenchman as DST agents. The printing restriction was later lifted. (CPJ)

On 31 July King Mohammed VI proposed the foundation of a royal institute to take account of cultural claims by the nation's Berber community. The institute would seek to introduce Tamazight, the Berber language, into the state education system. (IHT)

## MOZAMBIQUE

A CPJ delegation visiting Mozambique in July found the situation still tense even eight months after the murder of investigative journalist **Carlos Cardoso** (*Index* 1/2001, 3/2001). Mozambican journalists told the delegation that they are afraid to cover sensitive stories, particularly those involving corruption. The country is known for an environment in which both independent and state-sponsored media have competed freely, without official interference, noted the CPJ, making this kind of self-censorship especially disappointing. Cardoso was a veteran independent journalist who edited the daily fax newsletter *Metical*. He was shot dead on 22 November 2000; a week before he had launched a campaign against what he called the 'gangster faction' in the ruling Mozambique Liberation Front (FRELIMO), which he accused of provoking recent political violence in the country. (CPJ)

## NAURU

On 6 August AFP correspondent **Michael Field** was barred from covering the Pacific Forum Summit eight days before its start. The bar was linked to Field's investigation into money laundering in Nauru. US and Russian authorities claim that US$70 billion in Russian mafia money has been laundered via the island – where 400 offshore banks are registered to one government mailbox. (CPJ)

## NEPAL

On 15 June **Yubaraj Ghimere**, **Kailash Shirohiya** and **Binod Raj Gyawali** (*Index* 3/2001) were released on bail after spending ten days in detention. The three journalists were arrested on treason charges on 6 June following the publication of an article by Maoist guerrilla leader **Baburam Bhattarai.** (Center for Human Rights and Democratic Studies)

## NIGERIA

On 21 August, the Northern Christian Elders Forum (NORCEF) in northern Nigeria claimed that local churches were being destroyed as part of an Islamisation campaign supported by the region's state governments. NORCEF chairman **Adamu Baikie** claimed that churches were being destroyed, purportedly because they were built in residential zones. Some ten northern states have adopted the Islamic Sharia legal code, but Baikie says the laws are also being applied to non-Muslims. (*Guardian*).

**Nnamdi Onyenua**, editor of the weekly Lagos-based magazine *Glamour Trends,* was arrested on charges of criminal defamation on 8 June. Onyenua's article in the magazine's 6 June edition alleged that Nigerian president Olusegun Obasanjo was receiving US$1 million in allowances for each overseas trip, and that as of 30 May 2001 he had amassed US$58 million in allowances over two years. CPJ's sources said Onyenua was detained for more than 11 days pending investigation of the case, despite the fact that Nigerian law says no prisoner can be held more than 24 hours without formal charges. On or about 21 June, he was charged with publishing false information and defaming the president, and then released on bail. (CPJ, IFEX)

## PAKISTAN

On 4 June the Urdu-language daily *Mohasib*, published in Abbotabad, Northwest Frontier Province, was closed down indefinitely and four of its journalists were put under 'protective detention'. Four days later, following demonstrations by fundamentalist Islamic groups, police arrested editor **Muhammad Zaman**, managing editor **Shahid Chaudry**, news editor **Shakil Tahirkheli**, and sub-editor **Raja Muhammad Haroon** for blasphemy. The charges followed the publication on 29 May of an item titled *The Beard and Islam*, by the well-known poet and author **Jamil Yousaf**. The article criticised the position of Islamic fundamentalists who argue that a man without a beard cannot be a good Muslim. Following the arrests, the four men were reportedly beaten by police, Islamic activists have harassed the men's families and newspaper employees' homes raided by police. On 15 June, some imams in Abbotabad threatened journalists **Sardar Abrar Rashid**, **Mir Muhammed Awan** and **Syed Kosar Naqvi**, leaders of the local press associations, with 'the worst consequences' because of their support for *Mohasib*. Even though both the Federal Ministry of Religious Affairs and the Northwest Frontier Province Law Department have said the article in question was not blasphemous and the Inspector General of Police in the province has urged the local police to drop the case and release the journalists, the four men remain in custody. Author Yousaf is still in hiding. (CPJ, RSF)

**Rehmat Shah Afridi**, the owner and editor of *The Frontier Post*, was sentenced to death on 28 June for alleged drug-trafficking offences. Afridi was arrested in April 1999, only weeks after he had published reports that accused members of the Anti-Narcotics Force of themselves being involved in drug smuggling. Amnesty International and several other human rights groups have criticised the 'trumped up' nature of the charges against Afridi. (*Independent*)

**Hayat Ullah**, a correspondent for the Urdu-language daily *Ausaf* in Mirali, north Waziristan, is currently in hiding after officials in the region ordered his arrest in late July for reporting on clashes between local tribal groups. Regional authorities issued an arrest warrant under the Frontier Crimes Regulation (FCR), a legacy of British colonial rule that grants virtually unchecked powers to government administrators in tribal areas. Parties accused under the FCR are denied due process of law. The *Ausaf* office

in Mirali has also been closed by police, as has a newspaper distribution centre in Speenwam run by Ullah. (CPJ)

On 18 August Dr **Younus Shaikh** (*Index* 3/2001) was sentenced to death for blasphemy. Shaikh was accused last year by some of his medical students of blaspheming the Prophet Mohammed by claiming, among other things, that he was not a Muslim until the age of 40, and that his parents were not Muslims. Shaikh will appeal but faces several more months in prison before a decision is made. (*Guardian*)

## PALESTINE

Attacks on journalists covering the conflict in the Occupied Territories continued. On 20 June, the CPJ released a study of incidents of journalists wounded by Israeli live rounds or rubber-coated steel bullets that suggested that the journalists may have been deliberately targeted. **Hazem Bader** (*Index* 6/2000), a freelance cameraman working with AP TV in the West Bank city of Hebron, said his car came under fire on 26 June from an Israeli Defence Forces (IDF) outpost near the settlement of Tel Rumeida, about 500 metres away. The car was plastered with Arabic, Hebrew, and English stickers to clearly identify it as a press vehicle. 'It was an open and clear area,' Bader said. 'No one was moving in the area'. (CPJ)

On 26 July IDF chief of staff Lt-Gen. Shaul Mofaz 'reiterated the standing orders concerning the safeguarding of journalists in the Occupied Territories and called upon the army's commanders to strengthen the awareness of those orders throughout the ranks.' Yet, as widely feared, there were soon fatalities. On 31 July 26-year-old journalist **Mohammad al-Bishawi** of al-Najah Press Office and 25-year-old al-Quds photographer **Othman Katnani** were among the eight killed when an Israeli helicopter attacked the Islamic Studies Centre in Nablus. After their deaths the IFJ warned that independent media coverage of the Israeli-Palestinian conflict would become impossible if journalists continued to be among the targets of violence. Palestinian journalists are suffering a disproportionate amount of discrimination and violence, it added. According to IPI, of 102 violations of press freedom recorded up to 24 July this year, the Israelis perpetrated some 87%. (CPJ, IFJ, Al-Awda Right to Return Coalition)

On 31 July Palestinian Authority (PA) Information Minister Yasser Abd Rabbu reportedly requested al-Jazeera satellite TV not to repeat film of Hamas leader **Hussam Khader** criticising the Authority. Khader had asked angrily what PA leader Yasser Arafat was doing flying 'from one capital to another like Alice in Wonderland' while 'the Israelis are slaughtering us like sheep'. He reportedly described the PA as 'a bunch of thieves protected by 70,000 policemen'. The PA's Voice of Palestine radio accused Khader of 'serving Israeli interests'. (*Middle East International*)

On 13 August, **Tarek Abdel Jaber** and **Abdel Nasser Abdoun**, a reporter and cameraman for state-run Egyptian Television, were assaulted by an unidentified Israeli soldier at the Qalandia checkpoint between Jerusalem and the West Bank city of Ramallah as they filmed in the area. Abdoun captured the attack on video. The IDF spokesman's office called the incident 'wrong and completely unacceptable', but accused the journalists of 'provoking the soldier'. The soldier was given a 21-day prison sentence suspended for two years, said the IDF, and barred from commanding positions (CPJ).

## PAPUA NEW GUINEA

Papua New Guinea news media were threatened, journalists attacked and a television station vehicle burnt during anti-privatisation clashes in the capital, Port Moresby, on 28 June. Staff from the *Papua New Guinea Post-Courier*, *The National*, EM TV and the Religious Broadcasting Association, were threatened and attacked by both student protesters and police. Papua New Guinea Media Council president **Peter Aitsi** said: 'The media will not take sides and must be allowed to portray developments without fear of harassment or assault.' (PINA, IFEX)

On 29 August Papua New Guinea's National Broadcasting Corporation (NBC) news director **Joe Ealadona** was suspended for airing allegedly anti-government programmes on the state-owned radio network. NBC managing director

Kristoffa Ninkama alleged the news director threatened national security by allowing reports on a protest by armed soldiers against plans to cut back the size of the Papua New Guinea Defence Force. Ealadona was also alleged to have allowed to air a public affairs programme on protests by PNG university students and to have aired a live broadcast on the launching of a new Labour Party. (PINA, IFEX)

## PERU

Retired Peruvian police colonel and former Ancash region anti-terrorism chief Idorfo Cueva Retuerto brought charges of alleged injury and defamation against journalists **Jesus Alfonso Castigione Mendoza**, **Martin Gomez Arquino** and **Hugo Gonzalez Henostroza** on 8 May. The complaint stems from articles published in *Liberation* and *Caretas*, and broadcasts on Radio Alpamayo, accusing Retuerto of physical and psychological torture on journalists and over a thousand innocent civilians during the years of the Fujimori/ Montesinos regime. (IPYS, RSF)

On 1 August an Internet company hosting a popular website calling for the extradition from Japan of disgraced ex-president Alberto Fujimori, www.devuelvanafujimori.com, was attacked by unknown web hackers dubbed '*ciberfujimoristas*'. (IPYS)

## PHILIPPINES

On 31 May **Joy Mortel**, a reporter for the *Mindoro Guardian,* was killed in Barangay Talabanhan. She was the fifth journalist killed in the country in the preceding six months. On 30 May **Candelario Cayona**, a commentator for Radio DXLL, was shot dead in Zamboanga City. Cayona had received death threats for his outspoken comments against local politicians, the military and Islamist separatist guerrillas. The other three victims are from Radio Mindanao Network (RMN). **Chito Acbang** was killed in late May, **Rolando Ureta** on 3 January (*Index* 2/2001) and **Olimpio Jalapit** on 17 November (*Index* 1/2001). Since the return of democracy in 1986, 37 Filipino journalists have been killed for apparent work-related motives. (CPJ, CNFR, WAN)

On 6 June a bomb attack targeted Radio DYHB in Bacolod City. The station is known for its reports on local crime and the fighting against the Abu Sayyaf Islamist separatists. The station had continued coverage despite a 29 May appeal by President Gloria Macapagal-Arroyo for a news blackout on the offensive against the guerrillas. According to local police the ingredients of the bomb would have been available only to the military. (CPJ)

On 14 June **Juan Pala**, a commentator for DXLL Radio Ukay of the University of Mindanao Broadcasting Network, was ambushed in his car in Davao City. Pala is in a stable condition and resumed airing his radio programme the day afterwards from hospital. No motive or suspects have been identified. (CMFR)

## QATAR

On 6 June three men burst into the office of the daily *al-Watan,* and beat editor-in-chief **Ahmed Ali** unconscious. The three men arrested shortly afterwards were linked to the Energy Minister Abdullah al-Attiyah, whose proposal to levy charges for water and electricity Ali had recently opposed. (RSF, WAN, World Editors Forum)

## RUSSIA

Russian police arrested 14 protesters on 13 July, including two representatives of Reporters sans Frontières, **Alexandre Levy** and **Vincent Brossel**, who were demonstrating against the selection of Beijing as host city for the 2008 Olympic Games. The demonstrators were attempting to unfurl a banner in front of a meeting of the International Olympic Committee, when Russian OMON riot police arrested them. A Tibetan monk and several Russian human rights activists, including **Alexandre Podriabinek**, editor of the newspaper *Khronika-Express,* were among the first nine arrested. Five others were arrested a few minutes later and their passports confiscated. Two days earlier Moscow riot police had arrested six exiled Tibetans and **Maxim Marmur**, a photographer with AP. (RSF, IFEX)

On 26 July the Russian military imposed strict restrictions on journalists covering the ongoing conflict in Chechnya, requiring them to be accompanied by an official from the press service of the Interior

Ministry at all times. Titled '*Restrictions Designed to Reduce Negative Coverage of Chechnya*', the regulations further restrict media access to the region through cumbersome accreditation procedures and rules that make travel within Chechnya dependent on the permission of local officials. (RSF, IFEX)

On 30 August, the Lipetsk offices of the regional television station TVK were raided by armed men acting on behalf of Energuia, a shareholder company of the station which wants to force a change of management. Station director **Alexandre Lykov**, whose news team has been strongly critical of the governor of Lipetsk region, Oleg Korolev, said a court decision had previously barred Energuia from calling the shareholder's meeting needed to legitimise changes of management. The station's news programmes were not broadcast on 31 August. (RSF, IFEX)

## SENEGAL

On 10 July, **Alioune Fall**, editor-in-chief of the newspaper *Le Matin*, was detained for questioning by Senegal's Division of Criminal Investigations (DIC). The journalist had written an article about **Abatalib Samb**'s escape from the capital's Reubeus Central Prison. The article cited mistakes which led to the escape of several prisoners and quoted a source close to judicial circles. Fall spent the night at the DIC, where he was asked to reveal his sources. (WAJA, IFEX)

## SERBIA & MONTENEGRO

**Milan Pantic**, a crime correspondent for *Vecernje Novosti*, was murdered on 11 June outside his home in Jagodina, killed by a heavy blow to the back of his head with a blunt object. Family members claim that Pantic received phone threats prior to his murder concerning articles he had written on crime and corruption. (ANEM, CPJ)

Enraged supporters of former Yugoslav president Slobodan Milosevic attacked journalists at a rally in central Belgrade on 28 June, angry at local media coverage of Milosevic's extradition to the international war crimes tribunal in The Hague. Cameraman **Milos Petrovic** was knocked to the ground and Beta news service reporter **Suzana Rafailovic** was roughed up. **Petar Pavlovic**, a photographer with the local news agency Fonet, suffered a serious injury to the jaw. Shouting: 'You are to blame,' the furious protestors attacked anyone they recognised to be a journalist. (ANEM)

Some 15 participants in a 30 June gay pride parade and some otherwise uninvolved passers-by were attacked as members of the Obraz Patriotic Movement and Svetosavska Omladina (St Sava Youth) joined supporters of the local football club to break up the event. The event, organised by **Labris – Group for Lesbian Human Rights**, **Gayten-LGBT** (Lesbian, Gay, Bi-sexual and Transgendered) and the **Centre for Promotion and Development of LGBT Human Rights**, collapsed in the face of the attacks. A public debate at Belgrade University's Student Cultural Centre was abandoned as the hooligans stoned the building and attacked gays and lesbians. Journalists were also targeted. **Mileva Vukic**, a contributor to Radio B92's *Ritam Srca* programme, was assaulted, as was another B92 journalist, **Brankica Stankovic**. Members of Obraz and Svetosavska Omladina, among whom was a priest, Zarko Gavrilovic, accused B92 of being anti-Serb and hurled religious and nationalist insults at its journalists. (ANEM)

Belgrade Central Prison governor **Dragisa Blanusa** was removed from his post in mid-July in punishment for publishing his account of Slobodan Milosevic's days in his prison en route to The Netherlands and a date with The Hague war crimes tribunal. The book, *I Guarded Milosevic*, had been serialised in the Belgrade daily *Glas Javnosti* all that week. Blanusa denied overstepping his authority as charged. 'We all had a different mental image of that couple [Milosevic and wife Mira Markovic] for 13 years, and even before that. Yet during those 90 days I saw the couple for what they really are; the people should see it too.' (ANEM)

## SIERRA LEONE

A group of 19 newspaper publishers worried about regulators from the new state-funded Independent Media Commission warned that they would unite to defend themselves against censorship and possibly support the opposition National Unity Party. **Thomas Gbow**, editor of the controversial *Exclusive* newspaper, said: 'We want (President Ahmad Tejan) Kabbah and his SLPP to attempt to muzzle us. This is going to be his final lap.' (*Progress*)

## SLOVAKIA

**Ales Kratky**, a commentator with the Slovak daily *Novy Cas*, is being sued for publicly defaming the president, reported IPI on 26 July. He faces up to two years in prison if found guilty. Kratky's comments on the Slovak president's national address included the view that the speech was a 'report of a state of mind of an arrogant egomaniac'. (IPI)

## SOUTH AFRICA

Media and free expression groups, including members of the South African National Editors' Forum (SANEF), strongly criticised the use of subpoenas to force media personnel to give evidence in criminal trials during the summer. *Cape Times* photographer **Benny Gool** and *Die Burger* editor **Arie Roussouw** have been summoned as witnesses in a gangland murder trial, and journalists **Jaspreet Kindra** of the *Mail & Guardian* and freelance **Sam Sole** to give evidence against former Truth and Reconciliation Commission investigatory magistrate **Ashwin Singh**, accused of leaking information to the media. Mathatha Tsedu, chairperson of SANEF, said: 'People allow us into these situations because they know we will respect their confidentiality. If we are called to testify, then we may as well be police consultants.' The media already face tough strictures under Section 205 of the Criminal Procedure Act, currently used to seize journalists' equipment and compel them to reveal sources. In May authorities cited Section 205 in an attempt to force Gool to hand over pictures he took during the killing of Cape Town gangster Rashaad Staggie by vigilantes. (MISA, CPJ, IFEX)

South Africa's proposed Interception and Monitoring Bill, which would empower the country's police, the National Defence Force, the Intelligence Agency and the Secret Service to 'establish, equip, operate and maintain monitoring centres' also drew widespread criticism in August. If adopted, the legislation would allow the government to monitor electronic and cellular communication, in some cases without warrants, under the pretext of curbing organised crime. (MISA, IFEX)

## SRI LANKA

From early to mid-June **Dharmeratnam Sivaram**, the editor of the *TamilNet* news website, was accused in the state and independent media of being a spy for the banned Liberation Tigers of Tamil Eelam (LTTE). On 8 June the English-language daily *The Daily News* included Sivaram in a list of people identified as 'LTTE spies'. On 17 June, the Tamil-language daily *Thinakaran* and the independent Sinhala-language newspaper *Divaina* accused Sivaram of being an LTTE spy. No substantive evidence was provided to prove the allegations against Sivaram, who lives in Colombo with his family. (Free Media Movement, CPJ)

Two people were killed and 30 seriously injured when police fired tear gas and rubber bullets on a massive demonstration in Colombo on 19 July. The shootings took place when police tried to seal off the capital from tens of thousands of demonstrators who were protesting against the suspension of parliament for two months on 10 July. President Chandrika Kumaratunga suspended parliament just before her minority government was to face a vote of no confidence which it was expected to lose. (*Daily Telegraph, Lanka Academic*)

On 22 July the government warned local publications that they would be prosecuted for disseminating 'false statements' about the constitutional referendum scheduled for 21 August. The warning did not define what a 'false statement' could be, or specify a punishment, but it indicated that the government was considering reviving a 20-year-old censorship law (*Index* 4/2000, 5/2000). A couple of weeks after the warning the referendum was postponed indefinitely. (AP, *Lanka Academic*)

On 2 August the Supreme Court lifted a ban on **Prasanna Vithanage**'s award winning anti-war film *Pura Handa Kaluwar* (*Death on a Full Moon Day*). The three-judge bench ruled that the government had no powers to defer the screening of the film, had curtailed Vithanage's freedom of expression, and must allow the film to be screened in the country by 15 September. The state was also ordered to pay Vithange US$5,500 in compensation. The government had banned the film in July 2000 on the grounds that it would hurt morale (*Index* 5/2000). (BBC, AFP)

**Issue 3/01, Race Matters**
*Erratum* 'Index Index' (p117): *Sunday Times* correspondent **Marie Colvin** was not, as we stated, wounded by Indian Army fire in Sri Lanka on 16 April. Her published accounts attribute her injuries to shrapnel from rounds fired by Sri Lankan Army forces. The error was in the editing.

## SUDAN

On 17 August Sudanese professor **Bona Malwal**, a senior research fellow at Oxford University in Britain, filed a court action in Nairobi against the leader of the Sudan People's Liberation Army Colonel John Garang. He seeks damages for defamation and an injunction to stop the SPLA leader from threatening to kill him. He is taking similar action against Michael George Garang Deng, editor of *Update* magazine, a pro-SPLA publication, which, says Malwal, falsely linked him to the assassination of southern Sudanese leader **William Deng Nhial** in 1968. (NDIMA, IFEX)

Journalist **Faycal al-Baqer**, a contributor to several Sudanese newspapers, was arrested in Khartoum on 20 June and his computer and fax seized. No reason was given. (RSF)

## SWAZILAND

On 4 September, the government of Swaziland appealed against the country's 31 August High Court ruling, which allowed *The Guardian of Swaziland* newspaper to resume publishing after a four-month ban under Section 3 of the Proscribed Publications Act of 1968. The Act gives his office unlimited powers to ban or suspend publications that do not conform with 'Swazi morality and ideals'. The paper moved quickly to produce an online edition (www.theguardian.co.sz).

On 30 August Swazi senators attacked the Sunday edition of the *Times of Swaziland* over a pair of satirical articles which they claimed would incite the Swaziland population to hate the monarchy and challenge the rule of King Mswati III. Led by senator Masalekhaya Simelane, they took special exception to the description 'loyal servants of the King are called "Bootlickers" and "royal hangers on".'

## SYRIA

**Nizar Nayouf** (*Index* 6/1992, 8/1992, 10/1992, 10/1993, 6/1994, 6/1995, 5/1996, 3/1997, 1/1999, 4/1999, 4/2000, 2/2001, 3/2001) was released during the night of 21–22 June, having been detained for 24 hours just six weeks after being released from jail. The detention prevented Nayouf from attending an arranged news conference to pass on information about abuses by security forces. Nayouf said hooded men kidnapped him on a Damascus street on the way to his doctor, blindfolded him, beat him and made threats against him and his brother. Nayouf opined that the intelligence services orchestrated the abduction and that the president personally intervened when he heard what had happened. On 16 July Nayyouf was permitted to travel to Paris for medical treatment. (RSF, WAN)

On 23 July reporters were allowed an unprecedented but controlled visit to the city of Tadmor (Palmyra) to show them the locations of mass graves of prisoners killed in their cells in Tadmor prison in 1980. Other reports indicated that al-Jazeera TV journalist **Mohammad Abdullah** had been denied the return of his reporter ID card and was therefore unable to carry out journalistic duties, although the reasons were unknown. **Salwa Estwani** of the BBC and **Maher Shemitli** of AFP have been subjected to intimidation, it was reported on 23 August. (*Syrian Human Rights Commission*)

## TAJIKISTAN

On 12 July **Dodojon Atovulloyev**, editor-in-chief of the Tajik language newspaper *Charogi Ruz*, was released from custody after being detained by Russian police for a week. He was arrested at Moscow airport, en route from his home in exile in Germany to Uzbekistan, by police armed with an extradition request from the Tajik authorities. They accuse Atovulloyev of 'publicly slandering the president of Tajikistan' and 'incitement to national, racial, or religious enmity'. He had published an article in the Russian newspaper *Nezavisimaya Gazeta* in January entitled 'Why Tajik Leaders Don't File Tax Returns' and an article was published in June accusing Makhmadsaid Ubaidulloyev, mayor of Dushanbe and head of the upper house of Parliament, of being a drugs lord. (AI, *Eurasianet*, CPJ)

## THAILAND

Twenty-four editors and journalists from state-run ITV's news service were dismissed after protesting against a directive from Prime Minister Thaksin Shinawatra demanding that the station moderate criticism of him and his party. Critics said the directive was unconstitutional and breached rights to report without interference. (*South China Morning Post*)

## TOGO

On 5 June, **Lucien Messan**, editorial director of the opposition weekly *Le Combat du Peuple*, was given an 18-month sentence, six months of which were suspended, for expressing 'falsehood and the use of falsehood'. Local sources believe that Messan's arrest and conviction was designed to intimidate the press in the run-up to October parliamentary elections.

On 2 July, a group of 30 armed men allegedly entered *Le Combat du Peuple*'s printworks and seized plates and printed pages of the newspaper. Togo's Interior Minister, General Sizing Walla, said he acted to prevent publication of 'texts of such a nature as would threaten public order'. He was referring to an article that implicated the government in a murder plot against jailed former human rights minister **Harry Octavius**. (CPJ, IFEX, IPI)

## TUNISIA

Journalist and rights activist **Sihem Bensedrine** (*Index* 2/2000, 3/2000, 4/2000) was imprisoned after appearing on the London-based satellite TV channel al-Mustaquilla on 17 June. She spoke out against corruption in Tunisia and was immediately arrested on her return on 26 June. She was later released on 11 August but is expected to be called for trial later this year. (RSF, CPJ)

Legal proceedings were launched against dissident journalist **Taoufik Ben Brik**'s sister and brother-in-law on 20 June. The Tunis High Court summoned **Saïda Ben Brik** to answer a charge of 'mutual violence and participating in an altercation'. Her husband, **Khemaies Mejri**, was brought up on the same charge as well as 'undermining accepted standards of good behaviour' and 'disrespecting other people's property and being insulting'. The case was opened one month after Ben Brik announced his candidacy for the 2004 presidential election on al-Mustaquilla. (RSF, CPJ)

On 7 July four Tunisian prison wardens were sentenced to four years in prison for torturing former boxer **Ali Mansouri**. Held on a criminal charge, he was tortured and chained in his cell in a bid to break his protest hunger strike in March. In April he was hospitalised and both his legs had to be amputated. Startlingly, the court also ordered the Tunisian government authorities to pay 300,000 dinars (US$210,000) to Mansouri in compensation. The Tunisian Human Rights League said the judgement and award was unprecedented, the first time that the Tunisian state was being made to share responsibility for tortures committed in Tunisian jails. (*Cairo Times*)

## TURKEY

Lawyer **Fatma Karakas**, Women's Union representative **Nahide Kiliç**, journalist **Zeynep Ovayolu**, **Fatma Kara** and torture victim **Kamile Çigci**, all speakers at a conference on sexual assault and rape in custody held in Istanbul in June 2000 are being charged under anti-terrorism laws. The charges are reportedly based on their use of the words 'Kurdish women'. (Evrensel-TIHV)

The play *A Beautiful-Ugly King* finally got a performance in the south-eastern city of Diyarbakir on 5 June after the authorities rejected requests to perform it in 27 different provinces. The theatre company had to apply to the local administrative court to get the local governor's ban lifted. Police later tried to prevent the sale of tickets, forcing the relocation of the production from the city's State Theatre to an alternative venue. When company representative **Baris Özat** informed the police of the new location he was reportedly insulted and punched by police officers. (Evrensel-TIHV)

Sixteen people, including trades unionists, writers and actors, began their trial in a Turkish military court on 29 June, in connection with the publication of the book *Freedom of Expression 2000*, which contains 60 previously banned articles. The defendants were charged under Article 162 of the Turkish Criminal Code (TCK), which bars republication of previously banned material. The charged were: **Cengiz Bektas**, architect, writer and president of the Turkish Writers' Union; **Sadik Dasdogen**, owner of BERDAN print house; **Yilmaz Ensaroglu**, president of the Islamic human rights group MAZLUM-DER; **Siyami Erdem**, 2000 president of the Confederation of Civil Servants' Unions (KESK); **Vahdettin Karabay**, 2000 president of the Confederation of Revolutionary Labour Unions (DISK); **Omer Madra**, director of Açik Radyo (Open Radio); journalist **Etyen Mahcupyan**; film and TV actress **Lale Mansur**; **Atilla Maras**, 2000 president of the Islamic Writers' Union (TYB); **Ali Nesin**, professor of mathematics and son of the late humorist and human rights activist **Aziz Nesin**; actress **Zuhal Olcay**; **Husnu Ondul**, president of the Turkish Human Rights Association (IHD); **Yavuz Önen**, president of the Turkish Human Rights Foundation and former president of the Union of Chambers of Engineers and Architects; writer and publisher **Erdal Öz**; **Salim Uslu**, president of the Confederation of Islamic Labour Unions (HAK-IS); **Sanar Yurdatapan**, composer and spokesperson for the Initiative Freedom of Expression. All faced a penalty of up to two years' imprisonment by the military court. The 16 were also separately charged by a State Security Court in connection with the publication of *Freedom of Expression 2000* under anti-terrorism laws for 'terroristic and Islamic propaganda', separately again in the Criminal Court of Istanbul-Uskudar, for insulting the state and once more, in the Assize Court of Istanbul-Uskudar, for insulting national symbols. A number of local and international observers were prevented from entering the military court on 29 June, including **Annie Pforzheimer** from the US Embassy, **Haaken Swane** from the Norwegian Embassy, **James Logan** from Amnesty International, **Eugene Schoulgin**, Chairman of International PEN – Writers in Prison Committee, **Louis Gentile** from Canadian PEN, **Nazif Öztürk**, President of Unification of Writers of Turkey and Professor **Veli Lök** and other representatives of the Human Rights Foundation. (Article 19, IFEX)

On 29 June, a State Security Court sentenced **Fikret Baskaya**, an academic and editorial writer, to a 16-month prison term for 'separatist propaganda by way of the press', following an editorial in the 1 June 1999 edition of the pro-Kurdish daily *Özgür Bakis*. (*Kurdistan Observer*)

The European Court of Human Rights ruled on 17 July that **Leyla Zana** and three fellow Kurdish members of the Turkish parliament had not been given a fair trial by the State Security Court that jailed them for 15 years in 1994. According to the ruling the Ankara court was not an 'independent and impartial tribunal' and had denied the MPs a fair trial under Article 6 of the European Convention on Human Rights. They were originally charged with treason, but eventually convicted for membership of the banned Kurdistan Workers Party (PKK). Zana and fellow Kurdish Democracy Party (DEP) MPs **Hatip Dicle**, **Orhan Dogan** and **Selim Sadak** have been jailed ever since. In their application to the European Court, the four MPs also complained that they were convicted for defending the views of the Kurds in Turkey and developing peaceful solutions to the conflict. The European Court unanimously held that the applicants' rights under Article 6 §3 (a) and (b) had been violated, in that they had not been informed in time of modifications to the charges against them and that they had not been able to have key witnesses questioned. It ordered Turkey to pay US$25,000 to each of the applicants for damages and an additional US$10,000 for costs and expenses. (IFEX, RSF, AFP, *Kurdistan Observer*)
The day after the European Court's ruling, the Turkish Constitutional Court broadened the scope of a December 2000 amnesty law, but refused to include charges of 'crimes against the state', levelled against political prisoners such as Zana and her colleagues and members of the PKK. The revised amnesty now covers

offences such as extortion of information by threats, escaping or complicity in escaping, arson and abuse of power by a government official. (*Cildekt*)

On 8 August 2001, the Turkish government's TV and radio regulators (RTÜK) ordered a ban on BBC and Deutsche Welle's Turkish-language news programmes. RTÜK president **Nuri Kayis** expressed his misgivings but said he was powerless to revoke the decision by the monitoring body's executive committee. He said he was going to appeal against the decision in the courts. The programmes targeted by the measure, as broadcast via the NTV news network, fell foul of audio-visual control laws barring regular or direct broadcast of foreign programmes in Turkey. Deutsche Welle immediately stopped broadcasts but the BBC is awaiting a court decision before deciding whether to act on the ban. (IFEX)

On 21 August Turkish officials banned and seized copies of journalist **Celal Baslangiç**'s book of testimonies linking the Turkish state to violent acts against Kurdish civilians in the country's south-east, *The Temple of Fear*. Baslangiç, a journalist from the centre-left daily *Radikal*, was charged with making 'unfair and false remarks against the military' and 'mockery and insults against the Turkish armed forces'. The book deals specifically with four operations against the civilian population conducted since 1989. He faces up to six years in prison. (RSF)

Turkish police seized patients' files from the Diyarbakir Representation of the Human Rights Foundation of Turkey (HRFT) on 7 September. 'There is a real risk that both the torture victims and their doctors will be exposed to harassment, arrests and further torture,' Amnesty International added. (AI)

## UKRAINE

**Oleg Breus**, publisher of the regional weekly *XXI Vek* in the provincial city of Luhansk, was shot dead outside his home on 24 June at point-blank range in front of his wife. The motive for the murder remains unclear. His death was followed by that of **Igor Aleksandrov**, director of an independent TV company in Slavyansk, in the Donetsk region of eastern Ukraine. He died on 7 July from injuries sustained four days earlier. The assailants attacked Aleksandrov at his studios, hitting him on the head with baseball bats. He died in hospital. Aleksandrov's colleagues believe the murder was connected to his television programme *Bez Retushi* (*Uncensorsed*) which featured investigative coverage of government corruption and organised crime. (CPJ, RSF, IFEX)

## UNITED STATES

*The Wind Done Gone*, **Alice Randall**'s parodic retelling of the 1936 saga *Gone With the Wind* from the perspective of Scarlett O'Hara's black half-sister, was finally cleared for publication on 28 June. A court ruled that an earlier ruling barring publication on grounds that it violated the original's copyright was rejected on First Amendment grounds. (*www.salon.com*)

On 18 July singer **Bonnie Raitt** and former Doors drummer **John Densmore** were arrested and charged with disorderly conduct together with other 20 activists for protesting outside Boise Cascade Corporation in Chicago (BCC). The protest was organised by Rainforest Action Network (RAN) which has been fighting BCC over its worldwide trade in old-growth rainforest wood. (*www.sonicnet.com*)

Freelance journalist **Vanessa Legget** was arrested in late July on contempt of court charges and faces up to 18 months in prison. Legget is writing a book about the 1997 murder of socialite Doris Angleton and she refused to hand over research materials, including tapes of an interview with murder suspect and victim's brother-in-law Roger Angleton to a federal grand jury. The Department of Justice has implied that due to Legget being as yet unpublished she is not a journalist and does not enjoy the right to protect her sources. (CPJ)

Colorado radio station KKMG-FM has been fined $7,000 by the Federal Communication Commission (FCC) for airing an edited version of the **Eminem** song *The Real Slim Shady*, while KBOO-FM in Oregon suffered the same fate for airing **Sarah Jones**'s *Your Revolution*. Both songs were deemed to contain 'unmistakable offensive sexual references . . . that appear intended to pander and shock'. The Eminem song had been edited first by his label Interscope and then by the radio in order to make it comply with FCC guidelines. But the FCC has recently issued new guidelines placing innuendo and context as factors in determining whether a song violates standards, regardless of the words. (*www.sonicnet.com, www.alternet.org*)

Environmental lawyer **Robert F Kennedy Jr** served 30 days in prison for trespassing on federal land while protesting against US Navy bombing exercises on the island of Vieques. While in prison his wife gave birth to his sixth child who has been named **Aidan Caomhan Vieques** to mark his protest. Several others have been prosecuted but the stiffest sentence was handed down to the island's mayor, **Damaso Serrano**, who was jailed for four months for the same crime. (*The Times*, BBC)

On 13 August Judge Ronald Lew granted photographer **Tom Forsythe** First Amendment rights to display his pictures of the doll Barbie in the face of efforts by Mattel, the doll's manufacturer, to prevent him. Mattel regards the pictures as crudely sexual and misogynistic and claimed infringement of intellectual property rights in a bid to ban Forsythe's pictures, which put Barbies in 'explicit' sexual positions, doing housework, roasting in the oven and impaled on fondue forks inside a boiling pot. Forsythe is the 65th artist sued by Mattel. (IHT)

Federal judges working San Francisco's ninth circuit court of appeal are in dispute with the court's administrative officer over monitoring software installed on their computers to detect the downloading of music, videos or pornography. The software, regarded as illegal and unethical by the judges, was recently unilaterally disabled in protest. Administration head **Leonidas Ralph Mecham** claims that disconnecting the software might allow security breaches while **Chief Judge Mary M Schroeder** said the software might violate the Electronic Communication Privacy Act barring unauthorised intercepts of electronic communications. (IHT)

An anti-smoking group based in San Francisco called Smoke Free Movie (SFM) is lobbying to classify films showing actors smoking as 'R', restricted to adults or accompanied children, in the same way as those containing sex or violence. SFM is also demanding that film companies run on-screen statements that no money has been accepted to show people smoking, make cigarette brands unidentifiable and run anti-smoking adverts before each film. (*Guardian*)

UK freelance photographer **Steve Morgan** was arrested while photographing a Greenpeace protest against the Star Wars missile test in Vandenberg US Air Force base. He spent six days in jail before being bailed on charges of conspiracy to violate a safety zone. If found guilty he faces up to 11 years in prison and a $255,000 fine. (*Press Gazette*)

Proposed legislation that could criminalise the disclosure of classified information has been re-presented by senator Richard Shelby. The original bill was vetoed by ex-president Bill Clinton, and includes sanctions against leaks of information not of national security interest and could be used to force journalists to reveal sources. (NYT)

## UZBEKISTAN

Jailed writer **Mamadali Makhmudov**, was reportedly transferred from Tashkent prison medical centre on 16 June, where he had been because of failing health, and sent to a strict regime prison in Chirchik. (WiPC, International Pen)

**Alo Hodjaev**, the respected editor of leading Russian-language daily *Tashkentskay Pravda*, was removed from the paper in July. He had previously organised an exhibition to promote awareness of state censorship, covering the venue walls with articles banned from publication by Uzbek authorities. (*Eurasianet*)

On 6 August prominent independent journalist **Shukhrat Badadjanov** failed to appear at a summons to a criminal investigation brought by the Tashkent Prosecutor's Office. The next morning international media reported that he had fled the country. Babadjanov, also a respected painter, was charged with forging a letter of reference by the artist **Ruzi Choriev** for his application to the Union of Artists of Uzbekistan in the early 1990s. (*Eurasianet*)

## VENEZUELA

On 12 June the Venezuelan Supreme Court rejected a petition filed by **Elías Santana**, coordinator of the civic group Queremos Elegir and columnist for the Caracas-based daily *El Nacional*. President Hugo Chávez criticised both Santana and Queremos Elegir during 27 August and 3 September 2000 broadcasts of his radio programme *Aló, Presidente*. Santana filed the petition to assert his right to reply on the programme as guaranteed by Venezuela's Constitution. The Court ruled, however, that the right of reply was intended for individuals without other access to a public forum, rather than media professionals with their own programmes and publications. But the Supreme Court's subsequent ruling on 14 June, number 1013, drew an even angrier response from the country's media. The ruling, according to media rights groups, reinforces state powers to charge journalists for vilifying public bodies or officials. The court held that that the media must avoid spreading 'false news or news that is manipulated', an open-ended demand that could, say IAPA and other media groups, be used to target the news media, journalists and any citizen who might criticise, contradict or oppose the government's actions'. (IAPA, IFEX)

## VIETNAM

On 17 May Catholic priest **Thadeus Nguyen Van Ly** was arrested in Hue Province allegedly for failing to abide by his probation conditions. Father Thadeus has spent ten years in prison accused of opposing the revolution and 'destroying the people's unity'. (AI)

## ZAMBIA

Zambian journalists felt the heat for critical press coverage of President Frederick Chiluba and his ruling Movement for Multiparty Democracy (MMD). On 25 July, **Bivan Saluseki**, a reporter for the independent Lusaka daily *Post* who wrote an article linking Chiluba to corruption, was arrested and charged with criminal defamation of the head of state. It carries a jail sentence of up to three years. On 2 August, police issued a 'warn and caution statement' to **Amos Malupenga**, the paper's deputy news editor, for an article which accused Chiluba of having 'stolen and shattered to pieces Zambia's dream'. The article quoted the criticisms of former deputy finance minister **Newton Ng'uni**, who was also 'warned and cautioned'.

Finally, on 17 August **Fred M'membe**, the *Post*'s much prosecuted editor-in-chief (*Index* 3/1999, 5/2000, 1/2001, 2/2001), was arrested and accused of defaming the president over an editorial entitled 'A Thief for President'. M'membe accused Chiluba of 'diverting money from legitimate purposes to his own personal use', asking, 'how are people supposed to deal with a lying and thieving president?'

Another long-time foe of Chiluba, the Lusaka-based Radio Phoenix, was shut down, allegedly because it had failed to renew its operating licence. Radio Phoenix has been banned and its staff harassed and intimidated by members of the security forces since 1997, when its studios were destroyed in a mystery fire. The IPI, among others, suspect that the actions against the station and the *Post* is part of a wider campaign to silence the independent media prior to elections later this year. Chiluba's government has tightened control of state-funded media under the newly appointed Minister of Information and Broadcasting, Vernon Mwaanga. Three weeks after his appointment, Mwaanga dissolved the boards of the state-funded *Zambia Daily Mail*, *Times of Zambia*, the Zambia National Broadcasting Corporation and the Zambia Printing Company. On 24 July two editors of the *Zambia Daily Mail* were suspended for six months after an article described Minister of Home Affairs Peter Machungwa, who was being investigated for corruption, as 'disgraced'. In late June Zambian newspapers warned that Mwaanga would be targeting the private sector too, saying he would 'not tolerate any nonsense from the private media from now onwards'. (MISA, CPJ, IPI, IFEX)

## ZIMBABWE

Zimbabwe's independent media entered a dangerous summer with news of a partisan committee set up to look into 'professional standards and journalistic ethics' in the country on 23 July. 'I think this is part of the wider exercise in which the state is attempting to limit media freedom ostensibly on the pretext of protecting the privacy of individuals,' said **Iden Wetherell**, editor of the *Zimbabwe Independent*. The same month the government invoked the Presidential Powers (Temporary Measures) Act to amend the Zimbabwe Broadcasting Corporation Act and the Broadcasting Services Act 2001 to form a new company that will control transmission services to broadcasters, as well as giving the state potential total control over Internet access. (MISA, IFEX)

On 14 August police arrested reporters **Lawrence Chikuwira** and **Mduduzi Mathuthu** from the *Daily News* Bulawayo office, in connection with an article that alleged that people had walked out on Vice-President Joseph Msika at a 'Heroes Day' gathering when he had called upon them to sing pro-government slogans. The two were released the same day. The next day, 15 August, four senior *Daily News* journalists, editor **Geoff Nyarota** (*Index* 5/2000) plus colleagues **John Gambanga**, **Bill Saidi** and **Sam Munyavi**, were arrested and charged with 'publishing false information likely to cause alarm and despondency in the public,' a crime under Section 50 (2)(a) of the apartheid-era Law and Order Maintenance Act (LOMA). The four men were released that same evening after a High Court judge ruled that detaining journalists under that clause of LOMA was unconstitutional. On 16 August police tried again, re-charging the four under a different clause, Section 44. This section makes it an offence 'to distribute or circulate any subversive statement among the public' and carries a prison term of up to five years. (MISA, IFEX)

On 22 August, **Mark Chavunduka**, editor of *The Standard*, was arrested over a story reprinted from the London *Sunday Times*, suggesting that President Robert Mugabe felt himself haunted by the ghost of the late commander of the Zanla guerillas, Josiah Tongogara. Chavunduka was charged with criminal defamation, but the police refused to name the complainant. The *Standard on Sunday* had earlier reported that Chavunduka, his news editor **Cornelius Nduna**, Geoffrey Nyarota, Iden Wetherell and **Basildon Peta**, special projects editor of *The Financial Gazette*, were on a 'death list' circulated by Mugabe sympathisers. (MISA, IFEX)

On 23 August Zimbabwe's Vice-President, Simon Muzenda, warned that the government would not hesitate to arrest journalists who wrote inaccurate stories about the country 'to please western detractors'. He was speaking at a Zimbabwe Republic Police gathering. On 27 August armed war veterans and supporters of the ruling ZANU-PF attacked *Daily News* reporter **Mduduzi Mathuthu** as he covered a confrontation at a farm outside Bulawayo. (MISA, IFEX)

*Compiled by: Andrew Kendle (India & subcontinent), Andrew Mingay (South America), Anna Lloyd (Central America & Caribbean), Ben Carrdus (China, Japan & Korea), Fabio Scarpello (North America, South-East Asia, Pacific & Australasia), Gill Newsham (Turkey), Haifa Hammami (Eastern Europe), James Badcock (North Africa), Katy Sheppard (Russia, Belarus, Ukraine & Baltics), Natasha Schmidt (Coordinator), Neil Sammonds (Gulf States & Middle East), Priya Mehta, Rohan Jayasekera (Editor), Ruairi Patterson (Britain & Ireland), Shifa Rahman (East Africa), Victoria Sams and Yasmine Gaspard (West Africa).*

Index Index is updated several times a week online ⇨ www.indexonline.org/indexindex

# Ukraine:

# Borderland

**In Ukrainian it's Kyiv, in Russian Kiev. As Ukraine celebrates ten years of independence, ancient histories have ignited linguistic and other differences that divide this buffer state between western Europe and the vast Eurasian subcontinent**

*Kiev, Ukraine: souvenirs on sale in the street. T-shirts for all occasions. Credit: Tim Smith*

# A matter of identity

We are not we, I am not I!

'But who, then, are you?'
*We* don't know –
Let the German speak!

The German would say: 'You are Mongols.'
Mongols, that is plain!
Yes, the naked grandchildren
Of golden Tamburlaine!
The German would say: 'You are Slavs.'
Slavs, yes, Slavs indeed!
Of great and glorious ancestors
The unworthy seed! ❑

*From* The Epistle *by Taras Shevchenko (1845). Translated by Vera Rich*

*Donetsk, eastern Ukraine: 'Barbie Doll Beauty Contest' in Lenin's shadow. Credit: Tim Smith*

6.

VERA RICH

# Who is Ukraine?

Ten years ago, when the Soviet Union fell apart, Ukraine and its northern neighbour, Belarus, began their new existence as sovereign states with a major advantage: unlike the other 13 'Union Republics', they each already had their own representation at the United Nations. Their presence was an anomaly – the result of a ploy by Stalin, which gave him, in effect, three votes out of the UN's original 15 founding members.

Nevertheless, the two republics enjoyed at least a formal presence in international diplomacy, and separate Ukrainian and Belarusan delegations existed in a number of UN-related bodies, where their activities could on occasion be more than a formal endorsement of Moscow's line – notably in UNESCO and, in the case of Ukraine, in the Danube Commission. Newly independent Ukraine and Belarus had no need to apply for membership of these bodies – all that was needed, formally speaking, was to notify the change of flag and state seal.

However, although international recognition is a necessary condition of statehood, it is by no means sufficient. For a country to survive – and in particular, for its citizens to be willing to endure the economic disarray and privation of the post-Soviet space – a sense of national identity is required. But the question of what that identity should be has by no means been resolved.

With a population of around 50 million, Ukraine is the most populous of the former Soviet republics, apart from Russia itself. However, that population is by no means homogeneous. There is a large minority who consider themselves ethnic Russians. There is, perhaps, an even larger group for whom, brought up and educated as Russian speakers until 1991, ethnic identity meant little more than an entry under 'Point 5' of their identity document ('internal passport'). There are various minority communities with historic roots on the territory of what is now Ukraine – Poles, Magyars, Romanians and Crimean

Tatars – who, having been deported under Stalin, have since 1988 been returning to Crimea and have settled there, often extra-legally. Paradoxically, the sense of Ukrainian identity is strongest in the region which, for close on 200 years, was separated from the Ukrainian heartland, Galicia, which, in the partition of the multi-nation state known as the 'Polish Commonwealth', was assigned to Austria, and thereafter, from 1920 to 1939, was incorporated into Poland.

'Nation-building' and 'state-building' have been watchwords of politicians both under Ukraine's first president, Leonid Kravchuk, and his successor, the incumbent Leonid Kuchma. Yet on the eve of the tenth anniversary of independence, an opinion poll indicated that, if a vote on Ukraine's independence were taken today, only 57–60% of the population would be in favour. A majority, true; but hardly one with which the state leadership can feel comfortable.

Nor is there any one obvious unifying foundation for 'nation-building'. Ukrainian is the sole 'state language' – but much of the population remains, in effect, monoglot Russophone. (This does not necessarily mean that some of them, particularly the more politically aware, may not consider themselves patriotic Ukrainians and strongly resent the fact that they had no chance to learn, as children, what ought to have been their native language.) Nevertheless, the language difference is potentially, and all to often actually, divisive.

With language, too, goes culture. Ukraine's national poet and 'prophet', Taras Shevchenko, wrote his poetry in Ukrainian. Though it is for his poetry for which Shevchenko is best known, his *Diary*, written with at least an eye to possible publication, and his autobiographical novel *The Artist* were written in Russian. The annual celebrations of Shevchenko anniversaries – his birth on 9 March 1814, his death on 10 March 1861 and the return of his body to Ukraine and burial at Kaniv on 21 May 1861, are state occasions, with the president laying the first wreath. The May solemnities, in fact, have as their motto words taken from the poet's *Testament*: 'in that great family'.

But how far and how deeply do the Russophone population feel themselves part of that 'family'? In May 1998, instead of holding the 'Great Family' celebrations at Shevchenko's burial mound in Kaniv, it was decided to hold them at Luhansk in eastern Ukraine. How far the Russophone – and at that time striking – miners of the Donbass appreciated the celebrations is an open question.

As throughout the former Soviet Union, religion has once again become a part of the social, as well as the personal, scene; but once again it is more a divisive than a unifying factor. Kyiv received its Christianity from Constantinople in the tenth century. However, in 1596, part of the Ukrainian Church declared for union with Rome, while still retaining its own hierarchy and the rites and forms of Byzantine worship. Religious dissension between the pro-Rome 'Uniates' and the remaining Orthodox fuelled the wars of the seventeenth century. Eventually, in 1839, the Uniate Church within the Russian Empire was, by order of Tsar Nicholas I, 'reunited with the Orthodox'. In Galicia, however, it survived until 1946, when a similar 'reunion' was decreed by Stalin.

During the late 1980s, under Mikhail Gorbachev's policies of '*glasnost-perestroika-demokratizasiya*', the Uniate Church re-emerged in western Ukraine, and its adherents tried to repossess their church buildings, which Stalin had assigned to the state-approved Orthodox Church, or else to secular use. There were protests, sit-ins and, on occasion, violent clashes between the rival believers. Then, after independence, came a split in the Orthodox Church in Ukraine – part of which chose to stay under the patriarchate of Moscow, while part established its own patriarchate of Kyiv. The latter argued, with some justification, that it has long been Orthodox practice for the Church to be self-governing (autocephalic) in any country where it commands the allegiance of the majority of believers, and that since Ukraine was now independent, it should have an independent Ukrainian Orthodox Church. The Moscow patriarchate would not acknowledge this claim; mutual anathemas and excommunications were proclaimed, while parishes and believers clashed over the right to church buildings.

To add to the confusion, there existed yet another Orthodox community – the Ukrainian Autocephalic Orthodox Church – set up after Ukraine's declaration of independence in 1918, and now existing only in diaspora. Some Orthodox parishes, disillusioned with all the local hierarchs, have now transferred their allegiance to that exiled Church. With three rival Orthodox and one Ukrainian Catholic (Uniate) Church contending for spiritual jurisdiction and the physical Church property, pleas from both presidents – Kravchuk and Kuchma – for church union as a foundation for nation/state-building, have fallen on deaf ears. The only progress so far has been a decision in principle for the Kyiv-Patriarchate Orthodox Church and the Ukrainian

*Kiev: nationalist rally under the statue of 'national' poet Taras Shevchenko. Credit: Tim Smith*

Autocephalic Orthodox Church to merge – with Patriarch Bartholomew of Constantinople, the *primus inter pares* of Orthodox hierarchs, brokering the deal.

Apart from this significant but small step, the contentions remain; with the ironic result that, although in Kyiv in particular, historic churches and shrines are being restored, and in some cases rebuilt completely, for the most part they remain in state hands as museums. Were the state to hand them back to one or other claimant, the inevitable result would be further violence.

All this at the 'mainstream' religious level – let alone such minority issues as the erection of wayside Orthodox crosses in the areas of Crimea once again settled by at least nominally Muslim Tatars.

These linguistic and religious divides are paralleled by the political spectrum. Western Ukraine looks towards central Europe, and has a living – if elderly – memory of pre-Soviet times under Polish rule. Its population is categorically opposed to any attempt to take Ukraine into a restored Soviet Union – and its politicians warn that in such a case they would secede. This is not an idle threat: theoretically at least, western Ukraine is large enough and economically developed enough to form a viable state, and includes Ukraine's 'second capital', Lviv. For the most part, eastern Ukraine looks towards Moscow. The government in Kyiv has to keep a balance. Its political discourse and desiderata suggest a reluctance to commit itself: close ties with the EU and NATO, but no thought of immediate membership – though application in the more distant future is 'not ruled out'; membership of the Confederation of Independent States (CIS), but non-participation in activities aimed at 'integration', and participation in the GUUAM economic association (Georgia, Ukraine, Uzbekistan, Azerbaijan, Moldova), widely viewed as a counterbalance to the Moscow-dominated 'Customs Union' of Belarus, Kazakstan, Kyrgyzstan, Russia and Tajikistan).

Nor are the ideals of nation/state-building served by the politicians who proclaim them. Ukrainians have little trust in their political leaders and legislators. The economic woes of the country are a fertile ground for corruption and cronyism of all kinds. Election to political office is believed to depend largely on 'administrative resources', in other words, who the aspiring candidate knows in the upper echelons of the state structures. Whether or not the general run of politicians and officials are, in fact, guilty of these offences, they are widely perceived to be. Media criticism of office-holders is routinely construed as libel, with fines imposed on both writers and publishers, and, on occasion, threats of closure. Sometimes there are more serious consequences – such as the disappearance of the whistle-blowing journalist Georgy Gongadze. Although his fate has never been conclusively confirmed, he is generally believed to have been murdered – and allegations that President Kuchma was implicated nearly led to the latter's impeachment.

Furthermore, the transition from planned to market economy has been disastrous for publishing; state subsidies have vanished, and serious literature – including the Ukrainian classics in which national consciousness should be rooted – has given way to the more saleable pulp fiction. The Union of Writers, the most prestigious literary body in Ukraine, has a membership largely drawn from an older generation. Its contribution to nation-building, consists chiefly in the inauguration of twin annual prizes for services to Ukrainian literature – the Taras Shevchenko Prize for such services within Ukraine, and the Ivan Franko Prize for work abroad. With literature becoming increasingly marginalised by society at large, and a younger generation of writers who see the Union of Writers as a fossil of the Soviet era, its influence is limited.

In his Independence Day speech this year, President Kuchma said that in spite of the undoubted importance of economic objectives, his

*Kiev: St Mikolai, Ukrainian Orthodox church newly restored to glory. Credit: Tim Smith*

first priority remains the 'intensive promotion and development of democracy and civil society', in which citizens can put 'civilised pressure on the authorities' and statesmen and politicians can 'live peacefully with the people, not meet them on the barricades'. Likewise, he stressed the need to adopt 'basic European standards and parameters for organisations in state and public life, while at the same time preserving national traditions'. Or, as Shevchenko put it in 1845 in *The Epistle* to Ukrainians 'living, dead and not yet born':

> Study, read and learn
> Thoroughly all foreign things –
> But do not shun your own!

What exactly, in today's context, Ukraine's 'own' things are remains to be seen. ❑

***Vera Rich*** *is a freelance writer and translator specialising in central and eastern Europe. In 1997 she was awarded the Ivan Franko Prize for 40 years' service to Ukrainian literature*

ANDREI KURKOV

# Ukraine's ethnic kitchen

**Language is a divisive issue in Ukraine's ethnic melting pot**

Kiev is a city that never hurries. It is already old and has taken its well-deserved place in history – indeed, in several histories, the most important of which are the history of Russia (*Kyivan Rus*) and the history of Ukraine. Kiev has, too, its Orthodox 'Vatican', the Monastery of the Caves to which, for centuries, pilgrims have made their way – in the Soviet era they were called 'tourists'. It has its great river, the Dniepr, in which the first Orthodox Christians drowned the pagan idols.

The historical conservatism of Kiev is unconquerable, and paganism lives on to this day – and has a greater influence on the daily existence and social life of the citizens than do the various versions of Orthodoxy. The idols and Scythian 'images' have evolved into enormous memorials and monuments which, like the idols of yore, still receive their 'sacrifices' – in the present, humane, days, flowers and wreathes with ceremonial ribbons of mourning. Each important monument had its day. Formerly, in the centre of Kiev, in the main street, the Kreshchatyk, there were two monuments to Lenin, the 'summer' and the 'winter' one. Flowers and wreathes were laid at both, irrespective of the time of year. The 'winter' Lenin wore an overcoat, the 'summer' one a light, three-piece suit. Lenin himself never visited this city, and did not like what he had heard of it: it lacked the revolutionary spirit. Perhaps that is why the new powers that be decided that two Lenins in one Kiev was altogether too much. One of them was sacrificed and, in its place, a new monument was erected. This is several times bigger than the Lenin monument and, as conceived by those who commissioned and constructed it, is intended to defend Kiev from Russia; or, in official and more politically correct parlance, simply to defend Ukrainian

*Kiev: Lenin in winter, abandoned in the yard behind the Kiev History Museum. Credit: Tim Smith*

independence from everyone and everything. High on a lofty Corinthian column stands a Ukrainian maiden with a garland in her hands, intended to represent the peace-loving nature of Ukraine. This monument was erected a few weeks back, and immediately acquired the nickname 'My second mother'. This nickname is interesting, in that it embodies the information that there is also a 'First Mother' in Kiev – the 'Mother Country'. It is the largest monument in Ukraine. An enormous titanium-alloy woman armed with a sword stands above the Dniepr and meets the trains arriving in Kiev from the direction of Russia. Add to this the fact that the nickname of the current Ukrainian president, Leonid Kuchma, is 'Papa', and we have a passable 'outdoor' family portrait.

The monuments of Kiev are a special subject to which I shall return. First, I should like to dwell a little on the rhythm of this city, of its sweet unhurried conservatism, which 90 years ago played a bad joke with the history of the Russian Empire. It was here, at the Kiev Opera House, that a student terrorist killed the only real reformer in the Tsarist government, Piotr Stolypin, who had tried to reform Russian agriculture. Many people believe that had he lived and carried through his reforms, the October Revolution would never have happened and Russia would have been gradually transformed into a normal European constitutional monarchy, and would, in all probability, have 'set free' its non-Orthodox colonies. What would have happened with the Orthodox colonies, I will not attempt to say. But in any case, it was here in Kiev that a revolutionary's bullet put a full stop to the bright future of the Russian Empire.

In those days, subjects of the Empire did not have their ethnicity entered in their identity documents. Instead of this 'fifth point' of the Soviet internal passports, the Imperial ones had 'religion', and the population of Kiev consisted of two-thirds 'Orthodox' Christians and one-third 'Jews'. The Orthodox went to church, the Jews to their synagogues. There were, of course, some Muslims, There were also Karaims – a small ethnic group of Crimean origin – and at the centre of Kiev they had their place of worship, the Kennassoy, built by the famous Kiev architect Vladimir Gorodetsky. All these ethnic groups lived together in peace and friendship. The only one that lived in compact communities were the Jews who, for the most part, dwelt in the Podol and Syrets quarters and in the area of today's Victory Square, formerly known as the *Yevbaz* (Jewish Bazaar). The other inhabitants of Kiev – Russians, Ukrainians, Armenians, etc. – lived in mixed communities and were known simply as 'citizens'. Ethnicity had no significance for them; Kiev and the communal life of a great city made them one. It was only later, in 1919 during the Civil War, that 'Ukrainians' came to Kiev in the shape of the bandit-anarchists of Semen Petlyura, clad in 'national' plumage.

They came and began the pogroms, first against the Jews and then others. And, in the eyes of Mikhail Bulgakov, who was living then at 13 Andreyevsky Spusk, they at once became 'typical Ukrainian nationalists'. Bulgakov remembers them as people who wanted to destroy the life of Kiev. He remembers and detests them. Until then,

Bulgakov had never given any thought to the origin of the people who lived together with him in Andreyevsky Spusk; one and all they loved their city and enjoyed their life there.

Until this life was cracked by Revolution, until someone blamed the Jews for causing this crack, and it became possible to attack them. That is the start of division by ethnicity. And it went further: into Russians, Ukrainians, Crimean Tatars. With the establishment of Soviet power, everything reverted to the past and, overflowing with enthusiasm for the task of building socialism, the question of ethnicity disappeared and was dissolved in common brotherhood.

Then came World War II, and multi-ethnic Kiev suffered a terrible blow. Virtually the entire Jewish population – about 200,000 people – were rounded up by the Nazis and shot at Babi Yar. After the war, anti-Semitism was the unofficial policy of both Stalin and Brezhnev; as a result, by the end of the twentieth century, most of the remaining Jews of Kiev were in Israel, Germany, the USA and elsewhere. The city where the famous Jewish writer Sholom Aleichem lived and worked for so long had lost its Jewish accent.

But before this, from 1945–46, in the centre of Kiev the majority of people spoke German. These were German prisoners of war rebuilding the Kreshchatyk. They made the Kreshchatyk solid and majestic and, it would seem, their mastery and talent gave the Kievans an architectural inferiority complex, which can be observed to this day; it is Turks and builders from western Ukraine who work on the most important buildings in the centre of the city. Kievan construction firms mainly work in the suburbs, building mundane apartment blocks.

There are cities where people are born and then move away for a better life (for example, Odessa, where a whole generation of Soviet writers were born in the 1920s–30s). There are cities to which people come. To Kiev they come for life. It has everything necessary for psychological and physical comfort. Within its limits of being southern, unhurried and provincial, it is welcoming. It is, of course, sluggish. This cannot help but have its influence on the current owner of the city, Ukraine. For ten years now, the latter has unhurriedly been awakening, trying to raise itself on its elbows after its long Soviet hibernation.

Any attempt to hurry things up, to change the rhythm of the city and the country, has run into terrible opposition. Here people would have to live to be 100 to accomplish in one lifetime what in more dynamic cities is achieved in 30–50 years. Here it has taken the embassy of The Netherlands six years to get permission to put up a new building. Here people would rather build something eternal and non-functional than the reverse: memorials, replicas of ancient cathedrals and monasteries. Here they consider it easier to build underground than on the surface. Here there is what has to be one of the most beautiful underground railways in the world, enormous marble halls under the city, under the Kiev hills.

The national (ethnic) question in Kiev does not stand up. It keeps being raised, but falls to the ground. To make the city go over to a single official language – Ukrainian – proved impossible. And although after ten years of Ukrainian independence only six out of 200 Russian-language schools remain and the rest have gone over to teaching in Ukrainian, this has not affected the language of the streets.

Some young, desperately 'nationally conscious' Ukrainians have recently opened a café in the government area of the city called 'The Last Barricade'; here Russian is certainly not recommended. It is hard to predict how long the café will last, and whether it will become a centre for young Ukrainophones. If it is true that the fate of something (no matter whether ship or café) is programmed into its name, then this last barricade will soon fall.

A seller who wants to prosper always speaks the language of the buyer, the customer. Hence in the bazaars of Kiev, Georgians and Azerbaijanis cry their raisins and dried apricots in Russian. Hence the professional beggars tell their tales – of burned-out houses, serious illnesses and the untimely death of parents – in *surzhik,* a popular mixture of Russian and Ukrainian. This is a living language spoken by almost everyone in the suburbs of large cities, and hated by the Ukrainophone intelligentsia. *Surzhik* may be called a parody of the Ukrainian language, but one can take a more sympathetic view; not only is *surzhik* a living, functioning and developing language, it reflects the particular culture of the people who speak it. It is the village come into the city with its traditions. It is a *sui generis* Masonic brotherhood, membership of which may be found at every level of the Ukrainian government and Ukrainian society. Because they help each other, they

*Kiev: fishing in the Dnipro River. Despite radioactive contamination from Chernobyl further upstream, many still eat or sell their catch. Credit: Tim Smith*

want to work together and serve people like themselves who speak the same language. And being aware of the competition in the national market, knowing the difference between the psychology of the people who speak *surzhik* and other Kievans, the professional beggars and cadgers have decided to speak in the language of their largest potential 'customer' or, perhaps more precisely, in the language of their victim – *surzhik*.

Kievan beggars earn in one day the average monthly salary of a teacher. Professional beggary and cadging returned to Kiev after the long hiatus of the Soviet era and at once became one of the more profitable forms of business for the 'managers' and team leaders of the cadgers. They bring in disabled people missing an arm or a leg from the provinces, take them to the suburbs of Kiev and establish them in private houses. The business is usually run by Gypsies who, every morning, take the disabled to work the busy spots and underpasses in the city, where they collect alms until evening. Then the team leader drives up, collects the takings and takes the invalids home to rest till next morning. Not long ago, in a private flat in Moscow, police arrested a group of disabled from Kiev, brought there by an enterprising team leader. Apparently, the professional beggars of Moscow saw this competition as a threat and tipped off the police.

Can a city be ethnicised, made monoglot or monoethnic? Certainly it can. But only by ethnic cleansing and massive punitive measures. Thank God, this cannot be done today in our country and it seems the current ethno-linguistic equilibrium of Kiev will be preserved. And Kiev, as of old, will speak to all in the language of stones, golden onion domes on the churches, and monuments. And, as of old, a monument to the eminent Ukrainian philosopher Hryhorij Skovoroda, who wrote his poetry in Latin and refused to take service at the court of Catherine the Great, will stand in the Jewish Podol. And opposite Kiev University, the monument erected back in tsarist times of the most famous Ukrainian in the world, Taras Shevchenko – who wrote his poetry in Ukrainian and his prose and diaries in Russian – will still stand.

When I was a child, there was a story doing the rounds among my contemporaries: a Japanese shows a Russian his clenched fist and asks, 'What have I got here?' The Russian doesn't know. The Japanese tells him, 'TV. Now guess: how many of them?'

This story came back to me quite recently when I went to a bookshop to buy a map of Ukraine. When I got back home, sitting in my armchair, I asked myself: 'And how many Ukraines are there on this map?' And I answered myself, 'At least two and a half.' Two and a half, that is, if you divide Ukraine ethno-linguistically. If you divide it socially or politically, you will get many more. But for me, as a man of letters, a person who earns his living by letters and words put together in novels and tales, it is first and foremost the ethnic-linguistic criteria that interest me. And they interest me first and foremost because as a writer (a trader in words), I write in my mother tongue, Russian. And today, the Russian language has no official status in Ukraine; Pushkin and Tolstoy are taught as foreign literature.

Throughout the ten years of independence, not a single statesman has set himself the task of creating a single nation, or uniting the citizens of Ukraine into a whole. The paradox is that when Ukraine was Soviet, it was more whole and monolithic than it is now. The present ethnic face-off is reflected in the policy and aims of individual but fairly influential political parties. If the Congress of Ukrainian Nationalists, UNA-UNSO, and other groups and parties come out in favour of enforced Ukrainianisation and the ousting of Russian from Ukraine, then the Communist Party of Ukraine, the socialists and even some centrist parties will come out openly in favour of giving Russian the status of a second official language. The Russian and Ukrainian languages have become blind weapons in a senseless political struggle and this means that the people for whom one or other language is their mother tongue will be drawn, actively or passively, into this struggle, into linguistic and political argument. In a country where out of almost 50 million inhabitants at least 12 million speak Russian, any kind of 'linguistic action' is, quite simply, dangerous. Dangerous, because it is much easier than carrying out economic reforms and creating a civil society in which all can feel they are citizens fully valued and with full rights.

Linguistic confrontation recently took on a completely new form in Crimea, to which some 200,000 Tatars, deported under Stalin, have returned. The *Mejlis* (parliament) of the Crimean Tatars has signed a document on cooperation with the National Movement of Ukraine – yet another party with a national/ethnic agenda. Now, in Russophone Crimea, Ukrainian fighters against Russian influence, and 'Russification' in general, have a remote-controlled 'weapon'. The Crimean Tatars do not, in any case, have the best of relations with the local government (mainly communists) which pays only lip service to the capital. The local government has its own interest in having enemies in the form of the Crimean Tatars.

It is not hard, then, to understand the logic of the Tatars in looking for support outside Crimea. It is all too easy to imagine what that support may lead to – a worsening of relations between Tatars and Russophones in Crimea.

Linguistic scholarship in Ukraine has gone even further by inventing the term 'titular nation': translated into the vernacular, this means 'being considered indigenous to that territory, the dominant nation shall have more rights'. The first hierarchical division of the nations of present-day Ukraine has already taken place: they have been divided into 'titular' and 'non-titular' – alien. It may seem 'normal' to Ukrainian adherents of nationalism to assimilate other ethnic groups. But in any normal society in a similar situation, it would be a matter of integrating the different groups into a single multinational and multicultural society. ❑

***Andrei Kurkov** is a writer and novelist. His most recent novel is* Death and the Penguin *(English translation, Harvill UK 2001)*

OLENA NIKOLAYENKO

# The most dangerous place

**Last year, Ukraine, along with Russia, became the most dangerous place in the world for journalists to work**

In June this year, just one month after launching his website *Kriminalnaya Ukraina* (www.cripo.com.ua), Oleg Yeltsov was questioned by Security Service agents. Articles from his series *From the Life of Derkatch Family,* examining business ventures of the former State Security Service (SBU) chief Leonid Derkatch and his son Andry, caught SBU attention. Without much ado, SBU began an investigation into a possible violation of Ukraine's criminal code for divulging classified information online.

The mechanics for muzzling journalists may differ from one case to another, but the outcome does not vary. After ten years of independence, Ukraine has virtually no independent press. President Leonid Kuchma is listed as one of the 'Ten Worst Enemies of the Press' in the 2001 report of the Committee to Protect Journalists. His reputation was considerably damaged by his alleged involvement in the murder of the opposition journalist Georgy Gongadze who disappeared on 16 September last year and whose death is still unresolved (*Index* 2/2001). Kuchma emerged from the scandal unscathed.

The development of Ukraine's post-independence media follows a pattern familiar in many countries in East and Central Europe – from the 'sweetness of unlimited freedom' to 'growing disillusion' as economic constraints limit its development and the state imposes legal and financial constraints. In an environment of progressively slower political and economic reform, the media struggled to liberate itself from government control.

The fragile economic climate hampered the rapid growth of new

*Lviv: intended as the new KGB headquarters, but now earmarked as the city's main tax office. Credit: Tim Smith*

media outlets. By 1993, inflation was up to 10,000%, advertising revenues were minimal, costs of production and distribution soared. As a result, many media outlets got into bed with business or political patrons who offered them financial support. In return, they sacrificed the luxury of independent opinion.

'The traditional forms of press finance – subscriptions, advertising revenues and state subsidies – were replaced by semi-official state support for those media that remained loyal to the authorities and unofficial support from shadowy, often illegal elements interested in PR,' said Nataliya Ischenko, a journalist at *Krymskoe Vremya*.

With the approach of national elections in 1999, interest in the media greatly increased. The Council of Ministers elected its four members to the National Broadcasting Council in March 1999, but Kuchma was in no hurry to appoint his four delegates, thus ensuring that the Council could grant no new broadcasting licences to his opponents. In addition, the cabinet issued a 'temporary' decree that increased the cost of radio frequencies ten times for the period running up to the election. Kuchma received more airtime on television in the final month of campaigning than all 13 of his challengers combined.

In all, the 1999 election campaign was a depressingly familiar saga of the incumbent who tightens his control over the media to retain power. The tax office's visit to the private television station STB and the subsequent freeze of its bank account two months before the elections was typical of the tactics employed.

And over everything hangs the threat of libel. Many newspapers learned to their cost how high certain individuals estimated their honour and dignity. It cost one journalist his life. Igor Alexandrov, a host on a regional television station in Donetsk, was sentenced to a five-year ban from working as a journalist and fined the equivalent of US$1,250 for libel by the Donetsk court. Oleksandre Lechinski, a parliamentary candidate, attributed the Central Commission's decision to annul his election result to Alexandrov's reports in March 1998. Alexandrov had referred to him as the 'uncrowned king of vodka sales in Donbass' and warned that 'the more Mr Lechinski's business grows, the more orphans and handicapped people we shall have in the country'. After the withdrawal of Lechinski's complaint, the case was closed in 2000. But the journalist's troubles were not over. On 3 July this year, Alexandrov, also director-general of the Slavyanks-based TOR television station, was attacked in the street. He died in hospital after over a week in a coma.

Self-censorship avoids such risks and is widely practised. 'We are afraid of publishing articles that may lead to a lawsuit,' says Nataliya Dyachenko, editor of *Den*. 'We are afraid of releasing any information unless we have cast-iron evidence.'

'The editorial policy of many publications is formulated under the pressure of financial and political groups that control them,' stated a report from the National University last year. The enigmatic disappearance of Georgy Gongadze put Ukraine's media in the spotlight for a few weeks last year, the year when, according to *Reporters sans frontières*, with four journalists killed apiece Russia and Ukraine replaced Sierra Leone as the most dangerous places to work.

Dmitry Kulikov referred to Oleg Yeltsov as Gongadze-2. How many Gongadzes does Ukraine need to sacrifice before it wins the battle for a free press? ❑

***Olena Nikolayenko**, from Ukraine, is currently a Muskie fellow at the Kennan Institute for Advanced Russian Studies, Woodrow Wilson Center*

**MYKOLA RIABCHUK**

# Two worlds and Big Brother

**Historical differences and the threat of Moscow are used by politicians to manipulate the population and retain their power**

When I came to the West for the first time, in 1990, I had serious trouble explaining to my hosts where I was from. I considered myself Ukrainian; my passport even had an entry to that effect. Officially, there was such a thing as the Ukrainian Soviet Socialist Republic with its own government, parliament and membership of the UN. So I'd say firmly: 'I'm from Ukraine.'

My interlocutors were unimpressed. 'Sorry?' the more polite ones asked. 'What?' others said, straining their erudition to its limits, 'Bahrain?' 'No,' I'd reply. 'U-krain.'

'What's that?'

'A republic of the Soviet Union.'

'Oh, Russia!' Americans would nod, happily. 'No!' I tried hard to be patient. 'Russia is also part of the Soviet Union.' The Americans were baffled. How could Russia be part of Russia?

Finally, I met someone who was unfazed by my explanation. On the contrary, he reacted like a professional:

'Which Ukraine? Russian or Polish?'

It was my turn to feel taken aback.

'Soviet,' I muttered. 'For the time being.'

Anyone who has travelled to both western and eastern Ukraine – Lviv and Donetsk, for example – will confirm that they seem to belong to different civilisations. Easterners and westerners speak different languages (Russian and Ukrainian), they vote differently, attend different Orthodox churches (in the east, religion is altogether weaker), adopt different cultural attitudes, and see the past and future of the country in a different way.

*Lwow, Poland (now Lviv, Ukraine), 1936: children working as station porters. Credit: Sikorsky Museum*

The confrontation between the two Ukraines is dramatic, for it is hard to reconcile Soviet and anti-Soviet historical narratives, or any notion of entry into the European Union, with the prospect of integration with the Russian-Belarusan Union. It is difficult to think in terms of a Ukrainian linguistic and cultural revival while accepting its marginalisation by the Russian language and Russian culture.

But this confrontation is ideological: the two Ukraines are abstractions rather than opposing geopolitical realities. It is easily said that Lviv represents one Ukraine and Donetsk another. But it is almost impossible to identify where one Ukraine ends and the other begins.

The further east and south you go the sparer the vestiges of 'Polish' Ukraine: the Catholic churches, the castles disappear. There are fewer

Orthodox churches too, and Ukrainian is less often spoken. This paradoxical link between Ukrainian and Polish culture (paradoxical because historically Poland never promoted either the Orthodox Church or Ukrainian culture and language) has led historian Yaroslav Hrycak to remark that Ukrainian national identity developed most fully in regions longest under the Polish partition – Galitsia or Volhynia, for example.

In the east, Russian Ukraine dominates – in a monstrously Sovietised form. The Donbas region has the highest number of murders, robberies, suicides, abortions, cases of VD, and the greatest drugs and alcohol problem in the country. Further west, the signs of Sovietisation diminish: fewer Stalinist blocks, fewer monuments to Lenin and streets named after him, decreased industrial pollution and less of the Russian language.

But even as far west as Lviv, Russian Ukraine makes its presence felt: Russian newspapers and books are sold everywhere for peanuts, low-grade Russian pop shrieks in cafés, and the business elite – like the criminal elite with which they are closely linked – tends to speak Russian, the language of the *nomenklatura* (members of the Soviet Communist Party, the Komsomol and the KGB) from which most new Ukrainians hail.

In Donetsk, the leading town in the east, it is considerably harder to see signs of the other, Ukrainian Ukraine. Its representatives live in the provinces and do not hold decent jobs in towns because a career used to be linked with compulsory Russification. They live in *kolkhoz* ghettos, supplying guest workers to big cities.

'Things have improved, though,' Ukrainian activist Volodymir Bondarenko says. 'We've opened a few Ukrainian schools [ten years ago there were none in Donetsk, a city with a population of 1.5 million] and classes are full. We are fighting to found our own newspapers and bookshops.' Don't the authorities help? 'Not much, but at least they don't stand in the way any more.'

Many Ukrainians feel this is one of the greatest achievements of the past decade. For the first time, they have their own state, which does not prevent them from sending their children to Ukrainian schools and does not persecute the Ukrainian language.

*Donetsk: a class for young Jews in the recently opened synagogue. Credit: Tim Smith*

The metaphor of the two Ukraines points to the geographical and ideological polarities symbolised by Lviv and Donetsk. At one extremity is Ukrainian, European Ukraine, looking to join NATO and the EU. At the other is Soviet, Eurasian Ukraine seeking union with the eastern Slavs, similar to Lukashenko's Belarus. The metaphor goes a long way to clarify the choice Ukraine has been trying and failing to make over the past decade.

During Gorbachev's *perestroika*, the more pragmatic sector of the *nomenklatura* took power here. In order to overcome their conservative rivals in the Party, and legitimise their own liberation from Russia, the *nomenklatura* made a tactical alliance with the opposition national democrats. The democrats were to take responsibility for the anthem, the flag and the ideology of the new state, including culture and education. The *nomenklatura* dealt with matters more concrete: privatisation, the transport of oil and gas from Russia, financial speculation. This transformed it quickly from a political class to a political-cum-business class: no longer *nomenklatura* but oligarchy.

Arguments over whether the democratic opposition could have done anything to prevent this continue to this day. Rukh (the largest opposition group) did not have enough support to take power and introduce radical political and economic change. It had the support of just 30% of the population which had voted for a non-communist, non-*nomenklatura* president in 1991. But over 60% supported Kravchuk – a sign that for Russian, bilingual Ukrainians, Vyacheslav Chornovil and Rukh were less a 'democratic' opposition than a nationalist one.

A tactical alliance between the *nomenklatura* and the national democrats was a necessity. Yet no political coalition or multiparty system developed. Over the past ten years, *nomenklatura* clans have been ruling the country fronted by former opposition leaders, compromising any idea of independence, democracy and market reform.

This unofficial 'ruling party' quickly sensed the weakness of its allies – their paranoid fear of selling out again to Russia. Moscow is a magic word that acts on Ukrainian democrats like nerve gas. 'Mind we don't capsize the boat,' the president's men say on all TV channels. Democrats agree: better these rulers than none. Better this president than a Russian one. It's hard to say how much real anxiety there is in this posture and how far it is an excuse for morally ambivalent collaboration with the regime.

The authoritarian system which broke down as the USSR collapsed revived in the second half of the 1990s throughout the post-Soviet space. Local councils, even at the highest level, are decorative. Real power lies with the presidential office and regional administrations under its control. These play a role similar to former Communist Party committees. Neither Ukraine nor Russia, let alone the central Asian states, functions by the rule of law. A phone call from the presidential office – as from the Central Committee under communism – carries more weight than the decision of any court. Hence the political and economic problems these countries face.

Robert DeLossa of the Ukrainian Institute at Harvard has dubbed this the 'blackmail state'. Everyone has to break the law to survive. Entrepreneurs evade tax because they cannot pay, tradespeople bribe the militia in public, workers produce things on the black and ordinary people go for years without paying for gas, electricity, water and heating because they can barely survive on what they earn. The 'blackmail state' serenely ignores these and other trivial (or indeed less trivial) offences. This lasts for as long as you stay loyal to the regime. Any attempt to rebel is punished instantly – in a way that is legal and formally vindicated.

The fate of the former Ukrainian prime minister, Pavlo Lazarenko, can serve to illustrate this. Over the years, Lazarenko had made a fortune worth millions. He also had the ear of the president, even though the parliamentary commission on organised crime had told Kuchma that Lazarenko's interests were suspect. In 1997, the businessman turned prime minister was careless enough to say that in two years' time he would be standing for the presidency. He lost his job instantly and for the past three years has been awaiting trial in a US jail on charges including money laundering.

Lazarenko's story, and particularly the ruthless destruction of his media empire, was a prologue to the equally brutal presidential campaign of 1999. Regrettably, international organisations were silent. This encouraged the authorities to organise a farcical referendum on changes to the constitution intended to strengthen the president's authoritarian powers. The Council of Europe condemned this charade, however, and the Ukrainian parliament refused to accept the changes proposed by the president.

Kuchma's administration was further compromised by publicity surrounding tape-recorded conversations allegedly between the president and his closest associates on the 'removal' of a troublesome journalist,

Georgy Gongadze, later found dead. Despite the outrageous content of these tapes – viewed by most people as authentic – neither the communist left nor the national democrats demanded the president's dismissal.

His unpopular regime survived the scandal with few losses, much as two years ago the president – who is equally unpopular – won an election simply by eliminating his opponents. The 'ruling party' manipulates the electoral sympathies of the two Ukraines and promotes itself as the lesser evil in both regions. Neither the pro-Soviet left nor the pro-western right have much liking for Kuchma and his oligarchy, but they like each other even less and fear that the other side might use Kuchma's fall to advantage.

Opinion polls show that about one-quarter of the Ukrainian population supports a Ukrainian, European Ukraine. About one-third supports the idea of a Soviet Eurasian Ukraine. The rest, a little less than half the population, say they don't know, are not interested or haven't decided. They may talk about integration with Europe together with Russia, about the free market with every conceivable guarantee of social welfare, about the revival of Ukrainian language and culture alongside continuing Russian cultural and linguistic domination. Some observers see this as post-Soviet schizophrenia, the symptom of a public consciousness traumatised by totalitarianism; others as a comic leap from a pre-modern to a post-modern age.

Ukraine's future rests with this third, undecided sector of the country, manipulated by the 'ruling party'. This silent, politically invisible, apparently absent part of Ukraine is the main political standby of the governing oligarchy, and a 'cold civil war' is being fought over it.

The party is happy with the status quo. It needs the two Ukraines in order to appear to be mediating between them in the eyes of its own citizens and of the world. But more than that, it needs the third, undecided Ukraine because that is its electorate, its way to quasi-democratic legitimacy. That is why it tries to ensure that people remain apathetic, ignorant, indifferent and fearful of any social instability. That is why the press and television controlled by the regime promotes conformism and relativism of the most cynical kind.

'I'm no angel,' the president says, by way of justifying his penchant for talking dirty. The media faithfully pick up the theme: 'Who is an angel? And who would replace him? That appalling communist? That

*Dnepropetrovsk, former site of a Soviet space and missile project: street swap-shop for scarce electronic parts.*
*Credit: Michael J O'Brien / Panos Pictures*

mad nationalist? You want anarchy? You want civil war?' Of course not. Nobody does. Better not rock the boat, it's none too steady anyway. Who knows, the next president might well be worse.

There are factors, however, which will prevent Ukraine from stagnating completely under this regime. Reforms, though inconsistent, have created conditions for entrepreneurial initiatives and given people economic independence. The education system has changed, as has the lifestyle of many Ukrainians. Society is more open, better informed and looks increasingly to western ways. Civic organisations including independent media, though confined to larger cities, are growing despite attempts by the authorities to bribe, marginalise or destroy them.

The oligarchic ruling party is not as monolithic as the Communist Party used to be. It is made up of competing clans. Beside coteries building capital from the re-export of Russian oil and gas, or engaged in shady financial transactions, there are new groups making money on the production of industrial or consumer goods. The flow of capital from Ukraine to banks on exotic islands has decreased, thanks to the policies of Prime Minister Viktor Yushchenko and to action against money laundering taken by western governments. This has encouraged the ruling class to invest in Ukraine.

*Nolens volens*, Ukraine is in Europe and too close to the borders of first world countries to be ignored. The West continues to overestimate the readiness of Ukrainian elites to submit to Moscow, as well as the capability of present-day Russia to take control of such a vast and complicated country. But it also underestimates the importance of contacts with the West for Ukrainian elites: the possibility of holding money in western banks, taking holidays in prestigious resorts, sending their children to the best universities and so on. The *nomenklatura* and the oligarchs did not restructure the country and bid farewell to the empire and communism to be stuck in their own gilded cage.

The West does not have to give way to Kyiv's threats of closer ties with Moscow, made as soon as embarrassing issues such as human rights, free elections or the media are raised. But it can and should introduce more effective control of monetary transactions and a more active visa policy refusing entry to politicians and businessmen suspected of corruption. As long as a visa to EU countries is a problem for a journalist or an academic but not for post-Soviet gangsters, discussions on western aid hardly look serious.

Ukraine needs help, but it is important to understand which of the two Ukraines deserves it more. A weak civil society, swamped by economic chaos and political violence? Or a corrupt authoritarian state concealing the old Soviet system under the insignia of democracy and the free market? ❑

***Mykola Riabchuk** is a Ukrainian writer and journalist, the deputy editor-in-chief of* Krytyka *magazine in Kyiv. In the 1970s he was a samizdat author and activist. One of his short stories of that period was published in* Index on Censorship *in 1991. Currently he is researching the post-Soviet mass media at the Institut fuer die Wissenschaften vom Menschen in Vienna.*
*Translated by Irena Maryniak from* Tygodnik Powszechny *(Poland), 2 September 2001*

ANDREI KURKOV

# Don't be afraid of the dark

### Outlet One

Electricity refugees are people who couldn't afford electricity at the new prices. You had to pay your electricity bill to the new owners of the supply companies; after the latest privatisation there were only five of these in the whole of the country. Prices shot up so much that anyone who could still afford to pay felt they were having to pay both for themselves and several hundred defaulters. The Electricity Lottery became popular: you could win anything from ten to a million kilowatts, and the top prize was having the arrears on your electricity bill cancelled. Nobody had won the top prize yet, and the winners of kilowatts couldn't benefit from their prize if they had arrears. People still bought the tickets, even though very few individuals won.

I didn't play the lottery. My apartment had already been confiscated to pay off part of my arrears. The one good thing about the situation was that spring had come, the warm Kiev spring of late April, so I should be able to get by in my worn jeans, light sweater and windcheater until the first frosts in October. Meanwhile, I had the thunderstorms of May ahead of me and the hunters behind me, as in some American movie. 'Persistent defaulters' didn't only have their apartments or houses seized; they were liable to a reinvented form of slavery. If the value of your house didn't cover your debts, a defaulter had to work for the electricity owners for several years.

I decided that wasn't on, and did a runner. I hid underground, and found hundreds of others hiding there just like me.

I met Marina in the dark, noticing her entirely by chance. She was wearing a watch with a luminous dial. Before I saw her face I knew the time on her watch, so all I asked was, 'Is that two in the afternoon or two in the morning?' A gentle voice replied, 'The only way to find out is to raise a manhole cover.'

I felt an immediate urge to lift one of the sewer manhole covers, and not only to find out whether it was presently day or night. I wanted us to lift it together so I would get to see her face and discover whether it was as sweet as her amazing voice.

'Let's do it!' I urged.

She agreed. She took me by the hand in the darkness and led me along, then stopped and let go of my hand. Her dress rustled beside my face and I realised she was climbing up the vertical well to a manhole. I groped and felt the bracket steps in front of me and I too climbed upwards. When I got to the top I noticed a small light-coloured knapsack on Marina's back.

The street lamps were lit in the street – the only free electricity in Kiev shed a feeble yellow light over the deserted street. That was the best night-time lighting the government had been able to squeeze out of the electricity company owners and people were afraid to walk out at night. Sometimes, flying like moths to a flame, they were caught by the private security police who checked their identity to find out whether they were in debt to the companies.

Seeing Marina's face, and especially her eyes, I was momentarily struck dumb. Her eyes were green and twinkling, and a fringe of light brown hair fell almost to her eyebrows. To the sides of her forehead the hair was long and covered her little ears as it fell downwards in waves to her shoulders and the white mohair shawl she wore over her lilac dress. I thought she must have been at the hairdresser's recently, so she couldn't be a long-term underground resident, and in any case she was dressed too well for living underground. She looked to be about eighteen.

'Have you been here long?' I asked.

'Almost a year. You mean the hair? No, I have my hair cut here.'

'Underground?' I asked in surprise.

'Yes,' she replied coolly. 'There are lots of hairdressers here.'

I passed a hand over my own disorderly hair which was also in need of a cut, but I had nothing to pay for it with.

'Why, do you want a haircut too?' she asked smiling. 'Don't worry. You can get a haircut down here for ten American cents.'

'And where would I find those?' I asked.

In reply she only smiled again.

Through the manhole opening the street lamps lit up a white arch flanked by pillars. It was the old entrance to the University botanical gardens from Lev Tolstoy Street. Behind it, the path wound away to Shevchenko Boulevard.

'I wish we could go for a walk in the park right now,' I dreamed aloud.

'Under the park!' Marina corrected me. 'Let's lower this manhole cover though, it's getting too heavy for me.'

Only then did I realise why the cast-iron cover had seemed so light.

**Outlet Two**

We had been walking five minutes or so along a wide sewer tunnel when a strange question suddenly came into my head: why did it smell so good here?

'It only stank in the old days,' her voice chimed in. 'In the old days they poured all sorts of shit down here, but the only factory still working in this district is the perfume factory. Breathe in. It takes you back to the smells of your childhood.'

I breathed in and, listening also to the muffled sound of her little heels, imagined myself in one of the avenues on St Vladimir's Mount. In the darkness my left hand found her right hand.

'Forgive me, I don't want to tread on your toes!' I immediately excused myself.

She did not take her hand away.

'In the light people can follow each other with their eyes, but in the darkness they need to use their hands,' she pronounced. 'Not far from here there's another manhole with a good view. Do you fancy a peep?'

I agreed readily.

Out of the next horizontal opening of the manhole we saw the night-time opera theatre with its centrepiece illuminated: the bust of the poet Shevchenko. I listened. It was quiet all around, only from somewhere far away came the sound of a police siren.

'How about climbing out and going for a walk?' I suggested, turning to Marina.

She shook her head.

'A couple of my friends went for a walk here. I don't know where they are now. There are safer places to take a walk. Let's go on.'

Having issued this instruction, however, Marina hesitated and took the knapsack from her back. She looked in, using the light from the street lamps, and smiled to herself. I wondered if she hadn't whispered something.

This time I took the main weight of the iron cover on myself and, waiting for Marina to go down, lowered it back into place by myself.

**Outlet Three**

In front of us, from round some underground corner, the weak beam of a pocket torch appeared.

'Turn right now,' Marina said.

The torch went out and several people passed us in the darkness. Somebody said, 'Hi!' They were carrying a radio which was turned on. 'Tomorrow will be warm and sunny,' the forecaster announced. 'The water temperature of the Dniepr will be 18 degrees and air temperature will reach a maximum of 28.'

I sighed.

'Missing the sunshine?'

'A bit.'

'If we're in luck,' she said, and the luminous dial of her watch rose towards her face, 'we'll see the dawning of tomorrow's sunny day.'

I smiled doubtfully. Luckily she couldn't see.

A torch beam was shone straight in my face and we stopped. The beam passed over us like someone's appraising glance, pausing for a moment on Marina's knapsack.

Startled by the unexpectedness, and also scared, I squeezed her hand and immediately felt a reassuring squeeze in response.

'Looking for work?' a man with the voice of a heavy smoker asked. 'Two hours up top, with security. Three *hryvens* each.'

'What do we have to do?' I asked.

'Protest.'

'What about?'

'What's it to you? You get given a placard and you hold it. You don't even have to read what's written on it.'

'We can't do that,' Marina's melodious voice replied softly.

The torchlight shining straight in my eyes was making them water. It was a powerful torch. I screwed my eyes up, the torch was suddenly turned off, and the hoarse voice muttered, 'Suit yourselves. I'm offering cash, not back-payments off your bills.'

His steps died away behind us and we walked on.

'They get on my tits!' Marina said heatedly. 'On the surface it's all those reps of Canadian wholesale companies, and here it's that lot!'

'Who are they?' I asked.

'Recruiters. Getting people for demonstrations. Did you want to go?'

'No, politics make me want to puke. Just thinking about a demonstration made it feel stuffy down here.'

'Let's have a breather, then,' Marina suggested. 'We're at *Kreshchatik* already. It's good here at night: no cars, the air is pure. Turn left here!'

We headed along a straight dark tunnel for about five minutes before Marina stopped. Her dress again rustled upwards, brushing against my face.

I went up behind her. Between us we shouldered the cast-iron manhole cover to one side. Marina took the knapsack from her shoulder, opened it for some reason, and placed it carefully in the roadway.

The deserted *Kreshchatik* was flooded with yellow electric light and seemed boundless. I pushed my head above the road level and looked round. The shop windows were brightly lit, displaying even at night their three-piece suits, their Adidas trainers, and bottles of gin and rum lined up like a military parade. To the left, above a modest entry, the familiar sign of 'George's' was lit. I used to drink coffee there. And over there to the right (I turned to look) they did good set meals which were cheap and included meat.

'Did you hear about the Chinese–Armenian restaurant war?' I asked Marina.

'The what?' she asked, surprise in her voice.

'You didn't then,' I nodded. 'Well, do you know this place?' I gestured towards the Caucasus Restaurant.

'Yes.'

'Even five years ago this restaurant was called the China Restaurant, with neon pictograms above it, and a big wall hanging of the Great Wall of China inside. It didn't last long. The Chinese owner refused to hand the restaurant over to the Armenians and was killed in a car accident. Now it's called the Caucasus and the pictograms have gone, but the Great Wall is still there inside, and the red lampions over the tables.

'Is it Armenian cuisine, then?' Marina asked, turning round.

I licked my lips. 'No, Georgian, but the cook's an Armenian. Anyway, let's move, this is making me hungry.'

I moved down one bracket step, waited for Marina to come down too, and then dragged the iron cover back in place.

'If you want something to eat,' Marina said thoughtfully, 'we can go to St Andrew's Defile. It's on our way. Mind you, it's always the same thing there, vodka, salmon and salted gherkins. Better than nothing.'

**Outlet Four**

Once more Marina and I were walking together through a tunnel of the sewers of Kiev. We were holding hands and even the warm foetid air wafting towards us on a level with my nose, oxygen-deficient as it was, could not dispel my feelings of romantic weightlessness. Our movements really were in slow motion, but that is not uncommon when you are walking in the dark. I liked that slowness. It had a fairy-tale quality. It was the way you should walk ahead to a joyous and, of course, radiant future. You would approach, see a closed door through whose rectangular frame bright light was filtering, you would take the heavy bronze handle, open the door, and be sent flying by a torrent of brilliant sunlight surging into the space newly liberated from darkness. It might make you go blind.

As we walked along I was eyeing Marina sideways. Amazing how quickly your eyes get used to the darkness.

Behind us we heard what sounded like either a sob or a smacking of lips. I turned quickly, just in time to see two points of light five metres away near the ground. Some animal, a cat or a large rat, shied back, turned tail and fled as if it thought we were predators.

'Wait here,' Marina said and let go of my hand. 'Here, hold the knapsack.'

I took it and threw my head back to look up and try to make out her legs which were quickly climbing the bracket steps of the entry well, which wasn't deep at this spot. The manhole cover clanked and let a piece of night sky through its opening. I screwed up my eyes and made out several stars and the edge of the moon, but a total eclipse followed immediately as the sewer cover was closed and it again became as dark as Egypt's night. I thought I felt something shift inside the knapsack.

'He's home,' Marina said, taking the knapsack and again holding my hand in hers. 'Let's go, we'll knock.'

She led me sideways.

'Put your head down and get down on your knees,' her gentle voice commanded, 'or you'll bump your head.'

We crawled along a low narrow passage which I could sense was rising gently. Marina crawled in front of me. She hitched up her lilac dress, which normally reached down to her knees, with her belt so that now it barely covered her hips. It was hard going behind her. It might have been better if I hadn't been able to see so well in the dark.

Three minutes or so later she stopped.

'What's your name?' she asked. 'I am going to need to introduce you.'

Had I really not told her my name? I must have completely lost my head.

'Vladislav, but everybody calls me Vlad.'

Marina nodded. I heard her give a muffled knock on a heavy sheet of metal. It was a coded knock, probably some short word in Morse code.

I heard a rasping as the metal slid aside, but no bright light fell on us, although it was marginally lighter on the other side than in the tunnel.

'Come on out,' a man's voice boomed. 'Are you alone?'

'No,' Marina answered, 'there are two of us. Vlad, come out.'

I climbed out into a small narrow room with a high ceiling. When I straightened up it was a good metre above me. Next to Marina was a tall man with a short crew haircut. He moved the heavy iron sheet back in place, then turned and looked at me with curiosity. He switched on the light: a weak 25-watt bulb dangling from the ceiling. I could now see that we had climbed straight out of a hole in the wall which was covered by what looked like a heavy damper. I leaned forward to examine it: it was as if we had climbed out of a stove built into the wall, a kind of hell. The knees of my jeans were black now, and my hands wet and dirty. I turned round.

The man with short grey hair gave me a friendly smile and held out his hand.

'Alexander Petrovich,' he introduced himself. 'Let's go through and warm ourselves.'

Marina and I followed him into the next room, which was also dark, and then we turned right into a small boxroom with a low square table in the middle. An old sofa stood to one side of the table, wedged between the walls, and there were folding chairs on the other side. A candle was burning on the table and lighting up a bottle of vodka next to it, a plate of sliced red fish, and a teacup with small freshly salted gherkins. In the corner, on another folding chair, sat a sturdy, younger man in jeans and a denim shirt.

'Sergei,' he introduced himself and immediately took a glass down from the high windowsill, filled it with vodka and passed it to me.

'Is this a basement?' I asked Sergei.

'This is no basement,' he said, slowly shaking his head. 'This is Gallery 36. There's a world of difference. Here, cheers!'

He emptied his glass and put it on the table. I saw that it had been made from a neatly sliced beer bottle of green glass. My own glass was stranger, ceramic, and with seven facets.

'Have something to eat with it. Don't be shy,' Sergei said in his deep voice.

Alexander Petrovich had meanwhile poured a glass for Marina, which she easily downed. She took a slice of the salmon and bit a piece off.

'How's your dad?' Alexander Petrovich asked.

'You think I see him?' Marina replied with a shrug.

The grey-haired man nodded understandingly, then looked at me and my glass. His hand went to the bottle and for a moment his gaze became heavy.

'Don't hold back!' he said, filling my curious china cup. 'Here's to our meeting! Look into my eyes as you drink. It's our custom here.'

I drank it, looking into his eyes. With the second glass I recognised the taste. It was King's Cup vodka, flavoured with medicinal herbs.

'You're lucky to have caught us!' Sergei boomed. 'We're just waiting for a customer.'

'Are we in the way?' Marina asked anxiously.

I noticed she hadn't got round to putting her dress down again. Her legs were amazingly slender, and now, in a normal relaxed human posture, she was diabolically attractive.

'Your dad looked in a couple of days ago,' Alexander Petrovich began again, his eyes fixed on the vodka. 'He bought a picture as a present for somebody's birthday or something. It was a good picture too, *Nocturnal Vessel* by Bludov. Nothing wrong with his taste.'

The mention of her father made Marina smile wryly. Alexander Petrovich noticed and stopped talking. He poured the third glass without talking either.

At this point someone tapped on the window. Sergei got up. He was quite tall too. He looked out of the window, then squeezed between me and the table and opened the iron entrance door.

'Herllo!' a distinctly foreign voice reached my ears.

A typical American with a typical American smile sat down beside us. He smiled the smile at each person at the table but did not introduce himself. He took an envelope from the inside pocket of his tweed jacket and put it on the table in front of Alexander Petrovich.

'Thurs urs a harf,' he pronounced with a strong accent. 'I'll bring the second harf tomorrow.'

Alexander Petrovich nodded. He took the envelope and threw it up on to the windowsill.

Sergei immediately got a clean sliced beer bottle down from the same place, filled it with vodka and placed it before the foreigner.

The American hit himself on the forehead, gave a smile of contrition and took a flat whisky bottle out of his side pocket. He put it on the table beside the candle.

'That will go nicely,' the grey-haired man nodded gravely.

I noticed Marina was looking worried.

'Alexander Petrovich,' she began. 'We'll be going. Only don't tell Dad I dropped by.'

We left Gallery 36 by the same route as we had entered it. I managed to take a look at several of the pictures hanging on the walls. In one a naked man was wheeling two naked women in a wheelbarrow.

Once back in the tunnel I straightened up and felt my head reeling slightly. The vodka had warmed my body and now I had to readapt to the subterranean microclimate. Marina loosened her belt and her dress slid back down to her knees. She tied the belt again, turned, and took my hand.

'You're not an electricity refugee, then?' I said, thinking about the conversation in the gallery, where her father had been mentioned several times.

'No, I'm a different kind of refugee.'

'What kind is that?'

'My father was the leader of the Social Democratic Party and I met a boy whose father was from a different party, the People's Democratic Party. To cut a long story short, I was forbidden to see him and was about to be sent off to study in England. So I ran away.'

'And the boy?'

'The boy was sent off to study in the United States. He didn't run away,' Marina said sadly.

'Romeo and Juliet all over again,' I joked, but the ensuing silence told me the joke had not been appreciated. 'I'm sorry. And this Alexander Petrovich?'

'Milozorov? He's the owner of the gallery. My father often buys pictures from him with party funds.'

'Enough said,' I said, and squeezed her hand.

The luminous dial of her watch was raised to Marina's face and a barely perceptible green light was reflected on her flat little nose.

'I hope I haven't offended her,' I thought.

But I said, 'Thanks for the gallery. We got a bite to eat and a few drinks out of it!'

She nodded. She carefully felt her knapsack.

**Outlet Five**

Now the tunnel was clearly descending and I guessed it was falling in parallel with St Andrew's Defile to Podol. A good place for us to meet the dawn.

I tried to make out the time on Marina's watch, but couldn't. I wanted to ask her, but at that moment she stopped and looked up. Voices were coming from somewhere above us.

Marina turned to face me and put a finger to her lips. In the silence the other people's voices seemed to be brought closer, and in addition to the voices we heard a metallic jingling. The voices were thin, either women's or children's.

'It's the monastery beggars sharing out their takings,' Marina explained, smiling. She was still listening carefully to the sounds coming from above, and looking around I saw that one of the low wells was immediately above us, so the beggars were counting their blessings on the iron lid of the sewer manhole.

'Where are we?' I asked.

'The entrance to St Frol's Monastery,' Marina whispered.

'I don't suppose you know anybody there, so we could warm up again and have something more to eat?'

'Not in there.'

We stood a few more minutes in silence, waiting for the voices of the beggars on the surface to become silent, and then went on. The distant sound of coins still jingled in my ears and I recalled the price of an 'underground' haircut. Ten American cents. It would be good to have not only a haircut but even to have it washed. But where?

**Outlet Six**

Marina stopped before a parting of the underground ways. Even in the darkness I could see the tunnel in front of us veering off in three directions. Music was coming from the right-hand opening.

Marina looked at her watch.

'Still too early,' she announced meditatively. She cast a glance upwards. 'Shall we have another breath of night air?'

I readily agreed. After the vodka and the salt fish I wanted water to drink, or at least some fresh air.

I went up first, bending my head so that the lid lay on my shoulders. I filled my lungs with the viscous subterranean air and, having raised it a little, pushed it over to one side. Immediately the moist night air blew agreeably in my face and my gaze was again drawn to the sky and to the few bright stars which make up the constellation of the Little Bear.

Marina came up and her hair brushed my nose. I sneezed and immediately bent down, so loudly did the sneeze echo in the deserted street. When the sound had died away I looked around to both sides.

'Where are we now?' I asked.

'The same place, Podol,' Marina replied levelly. 'Saviour Street. Over there was the Kiev Local History Museum,' she pointed to a grand villa which was visible behind a low little church.

'So what's there now?'

'The American ambassador lives there now,' she replied and suddenly bent her head down.

Her hand went unceremoniously to the top of my head and pushed it – I bent down too. At that moment a powerful searchlight beam slowly passed over the open manhole.

When we looked out again the searchlight was already probing the building opposite the ambassador's residence.

'Hey, why didn't you go to England, anyway?' I asked cautiously. 'Because of that boy?'

'Why are you so interested?' Marina narrowed her green eyes and look directly at me. 'What is it? Are you jealous of him?'

'What if I am? If a girl like you had refused to go to England because of me, I would . . .' I was slightly lost for words, and Marina giggled.

'You would what? You would have taken me off to America?' she asked laughingly.

'Why America? I would take you to the Crimea. There's no electricity there, and nobody who owns it either. Nobody's out trying to capture anyone else, it's warm all year round, there's the sea, palm trees . . .'

'The Crimea?' Marina repeated in surprise. 'That's an original idea. I'll think about that.'

**Outlet Seven**

While we were walking silently on I was thinking over my own words of ten minutes before. I thought about the Crimea, and the longer I thought the more I liked the idea too. After all, there are any number of abandoned houses, sanatoria, guest houses. You could find some comfortable room with a balcony and a sea view, make it warm enough for the winter, arrange it the way you wanted and live in peace and joy, following the Soviet soldier's adage: 'As far away from the officers and as close to the kitchen as possible'.

Something crunched under my foot and I paused for a moment. It was a throwaway plastic cup and beside it, by the tunnel wall, an empty wine bottle was gleaming dully. Feeling sportive I turned round and kicked the bottle and it clattered off into the darkness we had just come through.

'What are you doing?' Marina whispered, displeased. 'People are trying to sleep!'

'Where?' I asked in astonishment.

'Everywhere, and here too.'

I shrugged. We walked on in silence until our way was barred by a grille across the tunnel.

'Well, where did that come from?' Marina wondered in surprise. 'It wasn't here the day before yesterday. We'll have to divert.'

We retraced our steps three hundred metres or so and turned off into another slightly narrower tunnel. A broad cable ran along one of the walls of the tunnel rising at times to shoulder level, sometimes descending almost to the level of the cement floor. I had noticed mice running along it. For some reason their presence cheered me up.

'I hope you aren't afraid of mice,' I said to Marina.

'No, I'm only afraid of horses,' she replied in all seriousness. 'Let's look out here and see where we've got to.'

I swarmed up the well and with an already practised gesture slid the manhole cover aside. Marina's head immediately popped up beside me.

'Oh-oh,' she whispered. 'We've strayed from our route.'

'What route?' I asked, looking around. 'This is Contract Square!'

**Outlet Eight**

The next stage of our underground journey took forty minutes. We were walking along a straight tunnel in complete silence, apart from the squeaking of the mice who periodically overtook us along the cable.

'What's in the knapsack, then?' I asked when I got bored with not talking.

'Kittens,' Marina replied quietly. 'Three of them. My cat has had kittens.'

'So that's it!' I thought. I felt sorry for the three little kittens who seemed to be on their way to an unenviable fate when we finally reached our apparent destination, the River Dniepr.

'Why don't you just let them out here, underground? Look at all these mice running around,' I asked after a long pause.

'Somebody would crush them, they are quite helpless.'

The tunnel suddenly became narrower and lower. Now as we walked along our shoulders were almost touching the cable the mice were running along. Marina let go of my hand and walked in front.

About three minutes later we stopped.

Marina climbed only three steps upwards and tried to raise the manhole cover but couldn't. I rushed to her aid. Between us we heaved the iron lid aside and pushed our heads out. The air was already filling up with the translucent quality it has before dawn. Up above, the street lights and stars were shining. This time we were halfway between the lower end of the funicular railway and St Vladimir's Defile. I was just about to say something to Marina when I spotted two men busying themselves beside a Rolls-Royce with the customised registration plate MIKHA. They were about ten metres away from us.

'I don't suppose you know who Mikha is?' I whispered, glancing over at what was going on beside the classy car.

'Of course I do! He's the boss of the gas party.'

'What party?'

'Oh, I can't remember what it's called, but it's the party of the owners of the gas companies. But what are they doing there?' Marina looked closely at the two men who had raised the bonnet of the car.

'Put it close, between the battery and the instrument panel,' the cold voice of one of the men reached us. 'So the blast goes in not out.'

'Am I arguing?' the second replied, carefully lifting a black object the size of a shoebox from the ground and passing it to his accomplice. 'Just as long as the lead is long enough.'

'It is, it is.' The second put the object inside and something glinted in his hand.

'Here, take the screwdriver,' he said, handing the glinting object to his companion. 'That's it. Close it up.'

'Listen, they've put a bomb in that car!' I whispered to Marina. 'We've got to warn someone.'

'Who?' she asked in surprise.

'Well, this Mikha for one.'

'Take it easy, Vlad,' she said, stroking the knapsack she had taken from her shoulder with her hand. 'Up there on the surface they have their life to live, and down here we have ours. The two don't mix.. Let's get out of here.'

**Outlet Nine**

I walked on, feeling very depressed now. My thoughts of our eloping to the Crimea had evaporated. I was no longer bewitched by Marina. I could already see how this was all going to end: she would cast the helpless kittens into the waters of the Dniepr, and I would say goodbye to her and go back alone. I might just have to wander aimlessly in the tunnels until I came to somewhere I knew better than Podol, where I would be on home ground. Perhaps some day I would meet a likeable electricity refugee just like myself, and we would walk together beneath Kiev until better times came. Perhaps we would even go on those demonstrations to make a little money . . .

'Bend down!' the same gentle voice distracted me from my reflections.

I bent down and noticed that the tunnel had broadened. Then we turned right, walked another hundred metres or so, and Marina suddenly stopped and turned towards me. In the darkness I could make out a smile on her face. I made it out and was scared: I thought that now she was going to say goodbye and walk away, leaving me alone in this underground realm for ever. A shudder ran down my spine.

Her face came nearer and her lips touched mine. Only touched, mind, but I was shaken. There was something cruelly final about it.

'Are you leaving?' I asked in a whisper.

'What do you mean?' her warm breath seemed to touch my face. 'What about the Crimea?'

I swallowed my saliva. I felt a little easier.

Marina looked at her watch. She turned round. I heard a long familiar metallic sound. Marina unlocked doors invisible in the darkness. And suddenly those doors opened wide and before us was a square frame in which the Dniepr flowed, and a piece of the embankment strewn with plastic bags, old newspapers and other rubbish. The bright sun was rising over the river and seemed to have been painted with special paints of fire you couldn't look at for long.

Marina took the knapsack from her shoulder and straightened her white mohair shawl.

'Are you going to drown them?' I asked cautiously, glancing at the light knapsack.

Marina looked at me with sad surprise and shook her head.

'Let's go,' she said. 'We have very little time.'

We ran up the steps to the upper embankment which was deserted and desolate. Marina quickly walked over to the memorial to drowned sailors and turned, pointing to a small fountain in front of the memorial which wasn't turned on.

'Quick. Pick up the coins,' she ordered.

I climbed into the water and saw that the bottom of this small ritual pool was positively piled full of coins. I started picking them out and stuffing them into my pockets. Out of the corner of my eye I saw her pull the three kittens from the knapsack, two black and one ginger, and put them on the grass beside the memorial, in the warm rays of the rising sun.

'The foreigners will take them straight away. They love animals more than people,' Marina said. 'Perhaps they're right. Have you finished?'

We quickly returned to the lower embankment and stopped at the rusty door which had allowed us to see the dawn.

'Show me the coins!' Marina said.

I took the dosh out of my pocket and held it out to her.

'There, you see?' She took several small nickel discs. 'You've got enough for twenty haircuts now, if not more. Well then, shall we go and get a haircut?' She nodded towards the opening beyond which the familiar darkness of our Kievan underground realm awaited us. 'Or shall we go straight to the Crimea?'

In my mind realism struggled with romanticism and, for the time being, won.

I sighed and said, 'Let's get the haircut first.' ❑

***Andrei Kurkov*** *is a writer and novelist. His most recent novel is* Death and the Penguin *(English translation, Harvill UK 2001).*
*First English translation by Arch Tait © 2001* Index

# The shadow of Chernobyl

**In the face of government indifference, the victims of Chernobyl in Ukraine, Russia and Belarus still suffer the effects of the nuclear fallout from the explosion of the nuclear reactor in April 1986**

Fifteen years after the event, the closure of Chernobyl has changed nothing as far as the people affected by the 26 April 1968 explosion of the nuclear reactor are concerned. The radiation is still there. The caesium-137 deposited in the soil, in crops and in animal tissues is still poisoning the food chain, the River Dniepr that flows through Kiev is still radioactive, cases of thyroid cancer, particularly in children, may be as high as 11,000 according to a report from the Geneva-based UN Office for the Coordination of Humanitarian Affairs, and the lack of comprehensive epidemiological data is inhibiting any assessment of the full extent of the damage. As with the authorities in the USSR at the time of the explosion, the governments of the countries affected – Ukraine, Russia and Belarus – are anxious to play down the continuing consequences of the explosion and show little appetite for dealing with their victims. In Russia, the fate of the 600,000 liquidators from all over the Soviet Union – who were sent in to put out the fire and clean up the mess with no protective clothing and little information on what they were dealing with and its likely consequences – is all but forgotten. Jean-François Augereau writes in *Le Monde* that 'an international research project was able to arrive at an estimate of the average dose received by the 4,833 Estonian liquidators' (about 100 milliSieverts; in Europe, workers in the nuclear industry are advised not to exceed 20mSv). He adds: 'Of the Russian liquidators, 284,000 appeared on the Russian national medical and dosimetric register in 1992 . . . In January 1996 the same register only listed information on 168,000 of them: "contact had been lost with the remainder".'

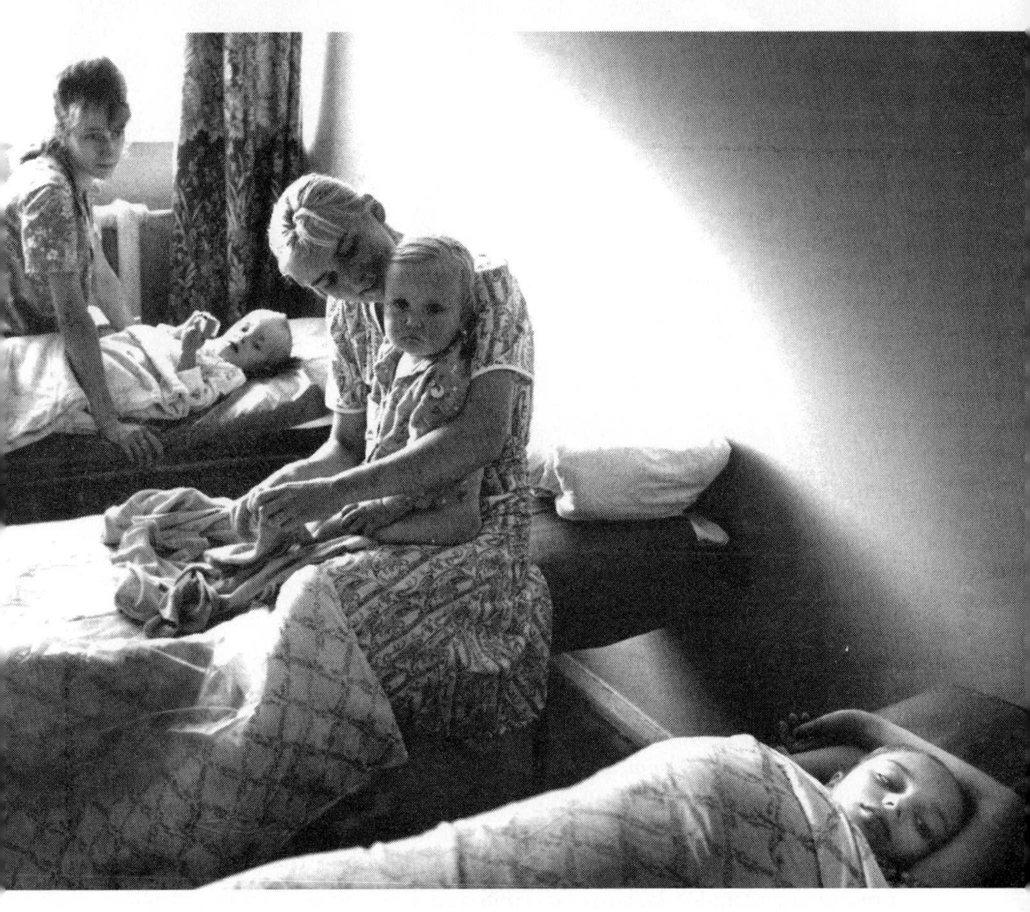

*Kiev Children's Hospital, 1992: leukaemia victims affected by Chernobyl fallout. Credit: Michael J O'Brien / Panos Pictures*

In Belarus, the country worst affected by the nuclear fallout from Chernobyl, one-fifth of the 10 million population, including 500,000 children, is living in heavily contaminated zones. Seventy per cent of the radionuclides thrown into the atmosphere by the fire fell on Belarus, as well as large quantities of the lead that was used to douse the flames. The government has expelled organisations like the Soros Foundation and the French-based Children of Chernobyl who might have helped. Ukraine, less severely affected, gets the bulk of the aid.

Nevertheless, work at the independent Belrad Institute in Minsk claims that 80% of the radiation affecting children can be traced to food. The biologist Yuri Bandazhevsky claims to have found a correlation between the amount of caesium-137 present in a child's body and the

*Above: Belarus exclusion zone: ruined and abandoned building in the contaminated area.*
*Right: Belarus exclusion zone: scrap metal scavenger collects radioactive metal for re-sale all over the country.*
*Credit for both pictures: Olivier Gachen*

incidence of heart, kidney and liver disorders. In 1999, Bandazhevsky was sacked from his post as director of the Gomel Medical Institute, imprisoned and then placed under house arrest.

Writing in *Le Monde* at the end of last year, Jean-Yves Nau claims that a report from the French Institute for Nuclear Safety and Protection (IPSN) stresses that no overall study of the development of the epidemic in all three countries has been published: 'We do have partial, preliminary data for Belarus for 1998 and 1999. It appears to confirm the rise in incidence and the continuation of the [thyroid cancer] epidemic in adolescents and young adults.' He adds that IPSN claims 'the real extent of the thyroid cancer epidemic is unknown, but it is still on the increase. Observers report increases ranging from area to area between a factor of ten and of a hundred times the natural incidence. This particular cancer is normally rare among children, but here the increase is especially high in the child population.' The IPSN concludes: 'Observation of this development tends to confirm that the thyroid cancer epidemic among children exposed to the Chernobyl accident seems to be continuing and to appear particularly among adolescents and young adults.' ❑

***JVH**, compiled from reports in* Le Monde, *16 December 2000*

**NATALIE NOUGAYRÈDE**

# Voices of Chernobyl

**Natasha (21), Russian, originally from Kazakhstan**, now lives in Svetilovichi, a village near Gomel in eastern Belarus. The Belarus government has helped tens of thousands of refugees from problem areas of the former Soviet Union to move into the contaminated zones. These displaced persons are used as a source of docile cheap labour. In an attempt to 'revitalise' the zone, the Belarus regime is also forcing trainee teachers and medical students to work for a year in the zone as a precondition for obtaining their qualifications.

'Quite a few Russian families came here from Kazakhstan, Tajikistan or Moldavia. In Semipalatinsk [Kazakhstan], where I come from, the problem is poverty. My parents and I sold the apartment and some of the furniture. We put everything we had into a container and spent nine days in a train to get here. A friend with contacts in Belarus suggested it. He said we'd be welcomed. When I went to the labour exchange in Gomel they gave me a job on a *sovkhoz* [state farm] that produces milk. They trained me for free to do artificial insemination of the cows. It looks as if it was some sort of shopkeeper who had our house before we did. He's left the area. When we first came, I was afraid of the radiation; I wouldn't touch vegetables from the garden. And then I thought, well, in Semipalatinsk [a former Soviet nuclear test area] it wouldn't necessarily have been that much different. Even so, I don't intend to stay long.'

*Credit for pictures pp190–96: Olivier Gachen*

**Yuri (64), a scientist,** lives within the 30km zone. He runs an 'Ecological Monitoring Department'. His team has no funding and almost no contact with the West, unlike their opposite numbers in the Ukraine.

'This lake is 2km long, and the radiation in the water is 3,000 times the permitted level. In Hiroshima, it was the atmosphere that was contaminated; here it all went into the soil. It will be 120 years before the soil becomes any safer. They brought me out of retirement to do this job. It's not so bad when you're old. We've lived our lives. Experts from the International Atomic Energy Authority came here. They left us a bit of decontamination equipment. They seemed mainly concerned about the risk of fire. I wonder whether the world has really grasped the fact that it's going to have to live with Chernobyl for at least another 120 years.'

**Petya (shown with his father Ivan), a worker** in the village of Bartholomevka in south-eastern Belarus, decided to stay put in spite of the evacuation measures. In Belarus, the official term 'exclusion zone' denotes areas within a few police checkpoints, which are easy to bypass. This covers only some of the regions that have been identified as dangerous and in which land is still being used. In Ukraine it is estimated that 5% of the land area was contaminated, but in Belarus, according to official figures, 23% of the country was affected. The population has no dosimeters that would enable them to check radiation levels in food (they're too expensive). The authorities claim that checks are carried out on the collective farms, a claim disputed by any employees you come across. In Belarus, the officially tolerated level of radiation in milk is twice the level allowed in Ukraine. There's no money to cleanse large areas of land. More and more the people affected by the crisis are feeding themselves by means of little family plots of land. They are slowly being poisoned.

'Twelve people stayed in the village. Down the bottom of the street there's old Misha with his paralysed leg. Over the road there's a field full of holes and bumps: that's where they buried some of the houses; they were radioactive. They had to stop them being looted. They left ours alone. What's the point in going off to die somewhere else? Round here there's a bit of game; you can go fishing; we cut wood to sell. We live like they do in the *taiga*.'

**Taïssa (8), leukaemia patient, is pictured in Minsk Hospital**. Her mother stands beside her bed showing a school photo in which the little girl is wearing a frilly dress and has her hair up. Chernobyl caused a huge increase in thyroid cancer in children. Other illnesses are on the increase, such as leukaemia; heart, liver and kidney disease; cataracts, and mental health problems. A Belarusan scientist, Professor Yuri Bandazhevsky, claimed to have established a correlation between the incorporation of caesium-137 into the organism of children (even in very small doses) and tissue changes in the vital organs. His reward for this discovery and for his denunciation of the waste of state aid to victims was to be imprisoned for six months in 1999. In October last year he was refused permission to respond to the European Parliament's invitation to come and present the results of his research. Contamination in children is primarily from food sources. Milk and beef concentrate caesium-137. Lack of controls on food produced in contaminated zones and distribution of these foods through the school meals service and through shops mean that there is no avoiding the risk. Nowhere is safe. Taïssa comes from Kobryn, in the west, a region at the opposite end of the country from Chernobyl.

'That's what she looked like. She always used to say: "My hair won't fall out." Natalia Ivanovna [her nurse] gave us a book and we read all about the side effects of the treatment. Taïssa was so sure of herself . . . but afterwards she had to give in and admit that she'd lose her hair as well.'

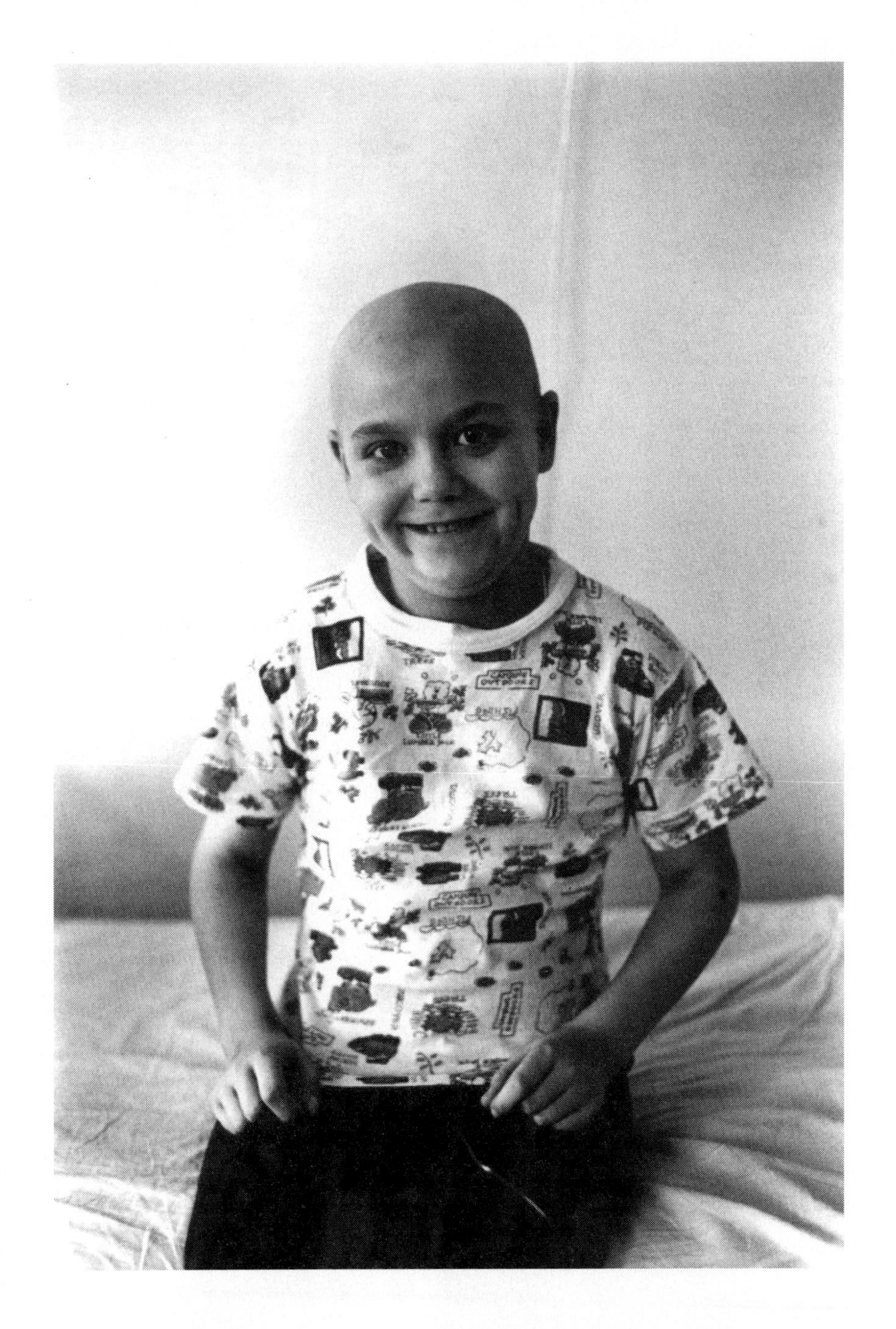

**Olya, a bakery worker, and her mother, Natalia, a pensioner**, visit the Bessed village cemetery, east of Gomel. The 120 inhabitants were evacuated in 1991, five years after the nuclear accident. Most of them were rehoused in Vietka and Gomel, on the edges of the zone.

'If anyone had told us that the village would disappear . . . In 1991 they cut off the electricity, closed the shop and blocked up the well to force us to move out. We were rehoused 10km away, in Vietka. We've come back today to our family graves; there's grandfather and two brothers. We're not doing anything wrong. Sometimes the police come and check. Why didn't they let the old people stay? People are falling ill and dying far away from their villages, but they still want to be buried here. We bring the bodies back . . . There are storks near the empty houses, but no people. The birds stayed behind though. The houses were looted; they took bricks away, some of them burned down. In Gomel, at the market, they sell berries and mushrooms that have been picked in the zone. If anyone decides to check the radiation levels in the produce, well, that just brings the price down, that's all.' ❑

***Natalie Nougayrède** is a journalist with* Le Monde, *France*

More articles, *Index* archive material and links ⇨ www.indexoncensorship.org/ukraine
Put your opinion online ⇨ www.indexoncensorship.org/comment
Email the editor ⇨ editor@indexonline.org

## MEHMED UZUN

# Culture: the nature of fiction

Mr President, members of the court: I would like to open my reply to the allegations against me by saying something about my profession as a writer. I think it apt to repeat what I said to thousands of my readers during a literary meeting in Diyabakir on 15 January 2000.

I am not in the service of any state, regime, ideology, political dogma, leader or organisation. I am simply a writer who feels the sorrow and pain of the oppressed and victimised; whose life revolves around literature and who seeks to create humane literature from this pain. Humanism, not ideology, not politics, informs my work – its language, style and narrative form.

I am concerned with human rights, minority rights, freedom of thought and the absolute freedom of literature. I do my utmost to protect these things, and my chosen form is through my writing. I have always made it a particular point to act in an open, democratic and civilised manner. This is one of the reasons I came from Sweden to attend this trial.

I am against all forms of terrorism, totalitarian ideology and anti-democratic action. Though I do defend the rights of the oppressed and victimised, I want to make it absolutely clear that I am opposed to violence every bit as much as I am to undemocratic behaviour. For me, proper relationships are characterised by civilised and democratic dialogue, a sense of ethical and moral responsibility, and the equality and freedom of individuals, religions and cultures. For everyone, everywhere, always.

And now, on to the allegations of the prosecutor. When my publisher first told me what these were, I said, 'No! It's not possible,' and told him I thought this was someone with a weird sense of humour; an April Fool's Day joke. However, by the time I had read the indictment, I saw

*Credit: Fredrik Funck/Pressens Bild*

this was no joke. I was shocked: with a stroke of his pen and in a few arbitrary sentences, the prosecutor had turned my novel *Daylight Like Love, Darkness Like Death,* on which I had worked with care for three years, into a simple propaganda tract directed at 'aiding and abetting a terrorist organisation'. This was horrific.

Under more congenial circumstances, I would like to have told the prosecutor something about my novel: about the thoughts, emotions, anxieties and desires that spurred it; my difficulties in writing it. This is not possible here. But allow me to make a few brief points. This novel is entirely fictitious, a product of the imagination. The 'Great Land' where a three-coloured national flag flies in the sky, which is surrounded by mountains to the north and the desert to the south, which is administered by a heavily moustachioed general, who took power by assassinating the King and viceroys of this former French colony, does not exist. There is no one called 'Kevok', the well-educated daughter of a good family who studied French language and literature at the university. There is no one called 'Baz' who had a tragic childhood and youth, grew up in orphanages, visits a loose woman who has grown old in the capital of the 'Big Country' – through which runs a river – who is a religious convert, sadist and sexual pervert. 'Mader', 'Jir', any of the heroes of the novel – none of them exists. The novel is pure fiction; the interpretation of time, space, names, place and all the rest is left to the imagination of the reader. For example, I was told numerous times by my Iraqi readers, who read the novel in the Kurdish original, that the country was Iraq; Syrians said that it was Syria; readers who read the novel in Turkish believed it was set in the Caucasus or Chechnya.

I have always responded to such comments by telling readers that the novel could have taken place anywhere where violence, terrorism and moral and ethical injustice prevail; and where people are in a straitjacket, desperate and with no way of escape. On the other hand, it could not have taken place anywhere where these factors did not exist. I had no political or ideological agenda. I wanted to convey the doubts and anxiety of an individual who has been paralysed by being trapped in an environment of violence and fear; to describe the fears and transformations; to describe, in fiction, the terrible silence and isolation of any individual who is prevented from expressing thoughts and feelings by the nature of the totalitarian society in which he is confined; and, finally, to highlight the virtues of tolerance and dialogue in the course of a narrative that would capture people's interest.

It is neither fear nor low cunning that led me to choose the novel form for my narrative. I have already written essays on democracy, the Kurdish problem and freedom of thought, for instance, in which I make my views absolutely clear. In this novel, I wanted to do something different in terms of language, style, technique, plot and concept. It was written in the closing years of a wild and murderous century. I wanted to turn to a blank page and welcome the new century with a novel that could be universally understood as a novel against rancour, hate, racism, nationalism, terrorism and totalitarianism.

The novel is my gift to the new century; the story of my profound hope that this century may be free of the murderous microbes of the last – rancour, hate, racism, nationalism, violence and terrorism; my desire that a common future characterised by tolerance, dialogue and democracy should prevail.

But my novel is not a treatise or a political essay; it is a work of literature that speaks of the human condition and universal values. The novel that is on trial has many reference points. I will mention three in particular: the epic of *Gilgamesh* is 'the first oral work written down for future generations of humanity' and tells of the search for immortality;

*Heart of Darkness* by Joseph Conrad, a British author of Polish origin, paints a literary panorama of Eurocentric racism and prejudice in the nineteenth century; *Light in August* by the US author William Faulkner tells in fictitious form of every tragic and sad event in the racist and conservative southern states of the USA. The title *Daylight Like Love, Darkness Like Death* refers loosely to these works. Light and love are two inevitable passions of human serenity. Everyone with a conscience should assist humanity to live in peace and serenity under conditions of freedom and equality and to approach one more step nearer the light. Everyone should contribute to the meeting of human beings beyond the evils of the last century by finding a constructive and enriching point of contact. It is strange that a novel with such profound human concerns should be condemned as a part of this rancour, hate and evil.

The prosecutor claims this novel is a piece of propaganda. I take this as an insult to my novel and authorship. I do not write propaganda for anyone and am nobody's instrument. My duty as a writer is to create a kind of literature beyond propaganda and at the service of everyone. I am a writer wed to a literary tradition who, despite everything, has not turned literature into a simple instrument of propaganda.

I want to give two small illustrations regarding this tradition. During that mad century that has just passed, a mad dictator called Hitler and his henchmen wanted to turn literature into just one more instrument of propaganda – as they had already done with so much else. Authors such as Thomas Mann, Hermann Broch, Elias Canetti, Robert Musil, Bertolt Brecht, Walter Benjamin, Theodor Adorno and Stefan Zweig, who defended the honour and respectability of the word and narrative, resisted, despite the many difficulties and sufferings this entailed.

At the same time, another megalomaniac dictator called Stalin and his lackeys did unbelievable things in order to turn literature into a simple eulogy to the great leader and a simple ideological tool in praise of the system. But Osip Mandelstam, Anna Akhmatova, Mikhail Bulgakov, Ivan Alekseevich Bunin, Boris Pasternak, Nina Berberova and thousands of other writers defended the value of literature even if it meant quitting writing and suffering in exile. It would have been grossly disrespectful to all these had I seen literature as a tool of propaganda and compounded it by writing this kind of stuff.

Literature as propaganda is bad. I would not deign to write this kind of literature even if the prosecutor wanted. I defend everyone's right to make propaganda as long as it is not an evil instrument of violence and does not hurt others; but I personally neither use the language of propaganda nor do I allow a sentiment that may be perceived as propaganda to seep into my literary work. The duty of good literature is to help man to continue to live and to retain hope, despite injustice and brute force, by reminding him of his humanity. It is not to serve as a weapon in political conflict. The purpose of literary work is always to recall and remind people of fundamental concerns, not to help them forget.

At one point in his indictment, the prosecutor says I shower some of my characters with praise. This is wrong also. I never praised anyone in this novel or in any other of my works. Eulogies do not make good literature; no more do invectives. Homer is our first great teacher on these things. Thousands of years ago, and just a few hundred kilometres away from where we sit today, there was a long and bloody conflict in the Dardanelles, in Troy. A blind narrator called Homer told of these painful conflicts in two epic stories, the *Odyssey* and the *Iliad*. There are thousands of heroes in Homer's literary gallery. But he does not praise or criticise them: he writes neither eulogies nor invectives. He simply creates a unique plot. If these two epics are still more effective than most modern stories, this is precisely why: Homer does not lampoon or praise, he narrates with intense human understanding. Neither Achilles' anger nor Hector's terrible despair represent the feelings of their creator.

Homer has taught us the art of narration. You must put a distance between yourself and the story. You must feel neither wrath nor enthusiasm, but will simply tell the tale and let events, emotions, your characters speak for themselves – and let the reader come to his or her own conclusions. You will never identify with the characters you create, simply give them life, spirit and a voice. And you will struggle as you write in the hope of ensuring that his life will be longer than yours.

What I am trying to say is that the heroes in my novel are products of the imagination and never aspects of myself. Neither Baz nor Kevok are Mehmed Uzun. Neither the anger, madness and wrath in the novel, nor the sadness, pain and despair, are mine. In fact, even the narrator, a foreigner who visits the country for a few weeks from Europe, is not me. If all these had been a reflection of Mehmed Uzun, we would have a stupid parody instead of a novel with serious literary and aesthetic concerns. The author has to see that his hero has a different fate from himself. Tolstoy created *Pilgrim Murat*, but the hero of the novel is not Tolstoy himself. Flaubert created *Madame Bovary*, but she is not the writer. If you want an example from Turkey, Yasar Kemal created *Ince Memed* [Memed My Hawk], but the life, fate and feelings of the novel's protagonist are very different from those of the writer.

The purpose and position of literature and the novel are different from the sciences. In 1949, William Faulkner wrote:

> Man is immortal, not because only he among life on earth has an inextinguishable voice, but because he has a heart, and there is mercy, sacrifice, patience and resilience in his soul. To write of these is the duty of a novelist.

Faulkner's words are a legacy for us. Our duty is not to occupy ourselves with what the prosecutor claims, but to recall human virtues such as compassion, mercy, honesty, pity, pride, sacrifice and solidarity, which are, in Faulkner's words, 'the troubles of the heart', with human warmth and purity. In a nutshell, the history of literature is a story of respect for humanity, the word and the narrative.

I wrote the novel which is on trial here in Kurdish, as I did my other novels. It was published in Kurdish in Istanbul in February 1998. *Daylight Like Love, Darkness Like Death* is the translation of the Kurdish original, *Roni mina evine, Tari mina mirine*. Kurdish, one of the ancient languages of Upper Mesopotamia that bears the mark of those civilisations, is my mother tongue. And Kurdish has been banned as the language of education and broadcasting for almost the entire history of the Turkish Republic [founded 1923]. Human beings do not have the freedom to choose their mother language but they are, none the less, responsible for it. I started to write Kurdish novels, despite the obstacles and prohibitions, because of an ethical and moral responsibility rather than any ideological or political motive. These prohibitions and the demeaning practices towards Kurdish are not right; I did what anyone with a conscience would have done. But I paid utmost attention to seeing that the novels and the art of the novel that I created in Kurdish were neither divisive nor damaging but uniting and enriching. As a polyglot writer – in Kurdish, Turkish and Swedish – who knows these cultures, my duty is not to put barriers between languages, cultures, literary traditions and countries, but to bring them closer to one another; it is not to increase prejudice but to reduce it. You can be sure that I will stop writing and will never write again if there is even a hint of rancour, hate, prejudice and violence in this novel or in any others. I know how cruel and meaningless these things can be.

This novel is neither dangerous nor harmful to the Turkish Republic nor to the Kurdish language, identity and culture. As history demonstrates, prohibitions of language, religion, identity and thought represent deeper moral and ethical blows to a country than economic crises and natural disasters. The prosecutor should be happy that this novel, which was written in Kurdish, has been translated into Turkish and gone through several reprints, and has reached so many readers. This, perhaps a first in the history of modern Turkey, has enriched the world of literature and culture. It is out of the question that the conspiracy suggested by the prosecutor should be lurking behind the wide readership of my novel.

This situation merely suggests that the civilised Turkish reader also likes a novel that is different and has been written in another language. I also want to point out that all the publishing houses that publish my novels are professional, and that I have nothing to do with the typesetting, layout, printing and distributing of my books.

Finally, in conclusion, let me say that the indictments of myself and my publisher are sad and meaningless. I want this situation to end not only for myself but for all of us. ❑

***Mehmed Uzun**, 04/04/2001 State Security Court no. 5, Istanbul*

# Migrating

Baz . . .
Kevok . . .

Kevok was killed. And Baz is taken to an unknown place.

Now we can begin our story from the beginning. Let's start with his childhood. His childhood years passed in bloodshed.

The story of Baz begins late one dark night when it was raining cats and dogs. In one of the interior regions of the Great Land, in some forgotten area, a place hidden behind mountains, protected by mountains, *Welatê Çiyan*, in the Land of Mountains . . .

Our story begins nearly forty-two years before the night when three bullets were emptied into Kevok's body, and she fell and lay on her back on the dry earth facing the moon and the stars.

Forty-two years before the night of her death . . . and five further years passed before these lines written in an attempt to describe Baz's life.

Let's now leave that terror-stricken night when death sealed her fate, and move on to a dark night long ago and arrive in a little village built on a high mountain slope. Baz is two or two and a half years old. A blinding darkness, pitch black, has settled over everywhere like a nightmare. A fine rain combs the hair of the forest, pouring the gift of plenty into the belly of the thousand-year-old earth. The dry earth sucks up the water, the earth breathes out freshness, the earth almost moans.

The owners who held this earth for thousands of years have abandoned their lands and villages. Tonight they are on the move, leaving their sorrowful hearts behind. Slowly, silently. With heavy hearts.

There's no remedy. They must leave and abandon their homelands. They must take refuge in the sheltering mountains and bottomless valleys. They must go to the eagles' nests and the lairs of wild beasts, it's their only hope.

Patiently, stubbornly, they have tested every way, but all ways are closed, they are going. Enemies they called strangers, who speak a foreign language, came in anger issuing an edict in a tongue they could not understand: they must go. They resist no longer, they cannot remain in their villages. They are helpless. Two ways lie before them, either death or the refuge of the mountain road . . .

The escape road.

They go. Children and the old on the back of horses and foals; their mattresses, blankets, mats, worn-out clothes, sacks of flour, wheat, barley, millet, rice and bulgur on the backs of mares; and on their donkeys, skins of oil, honey, grape syrup. They leave accompanied by a lament sung by wind and rain. They divide into four columns. The first column leads the way, guiding the big caravan, and consists of sharp-eyed marksmen, young fighters who will defend them against probable dangers. The second column includes women and children and the old, their past and their future both. The third column is the skilled workers and craftsmen, the middle-aged, the watchmen, the shepherds and herdsmen, driving before them, as closely as possible, their sheep and goats and cattle. In the last column are additional young fighters and marksmen.

Also in this last column go the elders of the tribe, the village soothsayers and storytellers, the spiritual leaders, fortune-tellers, interpreters of dreams, the carpenters, smiths, cobblers, farming experts and their wise men.

Their caravans set off; weak, wounded, lonely, defenceless, hopeless. All they want is to live, no more, only to live. In their own lands as of old. But now in order to live they must flee, to live they must give up their lands, must go far from their country, their land, their village and leave their graveyard behind. Not a sound from anyone, nobody speaks an unnecessary word, no one does anything superfluous. As though hands and feet were glued to the stirrup, as though their tongues were locked. Instead, the wind has a voice, the rain speaks. The rain falls in a drizzle, wind blows around them, horses whinnying, the mules gasp for breath, donkeys bray, sheep bleat, and the cattle plod on clumsily, lowing. Children cry, not understanding what's going on. The children who know nothing of country, land, migrating, time and fate, do not give up their childhood in this great migration. Their dogs and greyhounds are close to them. They haven't forgotten their oldest, most faithful friends.

Winter comes, sending news ahead with rain and wind; the dead of winter is on its way. Before winter arrives, the migrant caravan must reach somewhere with shelters to protect them from the fierce winter cold, from the wrath of dry cold, from attacks by wild beasts and foreigners.

For two years, yes, they had stood firm for exactly two years, sometimes running to the help of neighbours, sometimes setting traps where the foreigners passed, sometimes wounded and killed in fights. Now is the moment when all ways have given out and they flee. They know that the army has set off to chase them and will arrive in their villages before winter sets in.

One autumn day, the foreigners came with their tanks and artillery and guns. Why had they come here? No one knew the answer to this. They had just come like this, bringing violence and hardship. Why hardship, why violence? No one knew.

And the artillery exploded, the barrels of rifles and machine-guns spat fire, fire swept round the forests and villages. For hundreds of years silence had reigned in the Land of Mountains; in one moment it was destroyed and replaced by shouts and screams. Their screams mingled together with the shouts of the foreigners who didn't understand their language. The cries of the young men whose bodies were felled to earth by the pitiless bullets, the wails of old women robbed of their mourning and grief, the cries of mothers losing their children, the yelling of the young fighters as they leapt like hares and ran like greyhounds through the mountain streams, valleys and rivers, defending their own country against the raids of the foreigners, the bellowing of wild animals running to save their lives from the burning forests, the shrieks of birds as their fluttering wings tried to escape from the angry fire, everything mingled with everything. Screams and yells replaced the silence.

Despite all this, the real owners of the lands were hopeful. The land was theirs, handed down from their ancestors. Their ancestors' eyes had opened here, and here they had parted with life. They had owned these lands for seven, for seventy-seven generations. The bright moon in the skies, the hot sun, the trembling stars, the gently falling rain, the white, white snow, the majestic mountains, the fertile lands, the rivers flowing exuberantly and the eagles ever-present in the skies, were witness to this fact. This sky, these rivers, these icy lakes, these deep valleys were witness to their lives.

The deep voices still echoed in the deep valleys. The north winds still carried to the country the epics and the songs' melodies. They knew under what rock what wild beast has its lair, and on what high rock what bird makes its nest, and in what pasture what flower blooms. The country was their country, the land was their land. The foreigners would stay for a time and leave when they got fed up. That was what they thought at first, or rather what they hoped.

But they hoped in vain. Autumn, when the foreigners came, turned to winter, winter made way for spring, spring for summer and summer again turned to winter. The foreigners did not leave, in fact, on the contrary, they thrust in even closer. Everywhere their strange voices and foreign languages could be heard. They had armies, were expert in warfare, knew about tactics and had powerful weapons. Within two years, they controlled everything. They broke the villagers' resistance and cut off their breathing.

The mountain people were heroic; they resisted, strove, didn't give up till the end of the fight; but they were ignorant, lacked knowledge and experience, did not know how to take precautionary measures, did not know about tactics. They lived a divided life. They were split into households, families, lines, tribes and regions. They lived as of old, as they'd lived for thousands of years. They had no organisation, and no experience of fighting against an organised army, they were suffering from a shortage of weapons and ammunition, and had no knowledge of worlds outside their own. The traps they laid for the enemy were discovered immediately, the trenches they dug were quickly filled, the nets they wove were easily unravelled. The foreigners' methods of warfare had never been seen or heard of. This war was not like the wars with neighbouring tribes. They were defeated and the circle of fire surrounding all sides got tighter every passing day.

Now on this rainy night the foreigners were at the distance of one step. Or as close as one step away.

The two-year resistance was broken and the smell of gunpowder was taken over by the smell of blood. The smell of blood was everywhere. It penetrated the forest and the trees; the valleys and the stream beds turned into graveyards, corpses were strewn everywhere. The crimson blood dyed red the waters of the foaming white rivers. The houses were destroyed, the forests burned furiously. Gnarled gallows sprouted everywhere. Those who were captured were strung up on the gallows and the death sentences were read out to them in languages they could not understand. In a short time the foreigners had changed their lives which had run in the same courses for thousands of years and now they had forced this migration.

So it was the night of the migration. The villagers who were the real owners of these lands were going away from the villages. The first column had gone a long way and even reached the foothills of the mountain that stretched to the sky. The second and third column were going slowly. The fourth column was only just leaving the village. The last ones looked round at the villages before they left their houses for ever.

The houses in the darkness, the yards, the gardens, the stables, the wells in the village square, the poplar, walnut and fig trees were dejected, broken and deserted.

The lives here, the memories, weddings, feast days were entrusted to this rainy night and remained behind. They had opened their eyes to the world in these houses, made of black stone, they had got used to the pungent smell of animal dung in these sheep folds under their houses or those of their neighbours, in these fields that they were leaving behind they had planted corn, reaped the golden ears, harvested the crops. In this village square they had raced foals, thoroughbred Arab horses had galloped. In the mountains here they had hunted deer and gone hunting partridge and pheasant and birds of prey. The bears' moans had been heard on this mountain, the wolf packs had been hunted here. They had set traps on the fox paths and gone up this mountain for hunting falcon, hawk and eagle. They filled their children's bellies with the trout that they caught in the lake near the village.

In this village, which was blessed with no fortune, in the destitute nights, after the children had gone to sleep and the owls had started to hoot, they gave themselves up into the arms of intoxicating sexual desire and calmed the incessant urges of their bodies. On the hot summer nights in this village they had weddings and danced the *halay* and the round dance. The voices of the storytellers used to rise to the heights.

Now they were going, leaving behind everything they had lived through, their pasts, their memories.

There was the smell of dried dung, which they had left burning in the stoves of the houses they were leaving behind. The bewildered chickens, cats and dogs were wandering near the houses, not knowing where their owners had gone. Some of those who were leaving behind their houses, memories and lives were crying, others looked at the village for the last time and let out deep sighs.

Return? Would there be a return after this leaving? One day would they be able to return to their houses and old lives?

Yes, when the foreigners left. Then they would return to their lands. The villages would again be cheerful. The ashes in the stoves would be replaced by fire. The pots would boil on the flame, food would be cooked in them. The timid, silent moans of lovemaking would be heard at night.

The foreigners would go. What could they make of this mountain country, distant from the world? They would go, if not today – tomorrow, if not tomorrow – three days later. Sooner or later they would definitely go back to their big cities. their warm homes in their flats. They would be reunited with their loved ones – even foreigners have loved ones. They would go, those on the migration believed that.

Or rather hoped that.

Now they were passing the crop fields and pastures. There was the smell of wet earth. The fields had been recently harvested. The storerooms had been filled with grain. Now they were going away, leaving the winter provisions in the storerooms, as well as the fields.

It was raining, wetting the earth and washing everything clean. If one didn't count the occasional jackal's howls, only the noise of the rain accompanied the migratory caravan. It was dark, blind dark everywhere . . . ❑

*From* Daylight Like Love, Darkness Like Death
*Translated by Richard McKane and Ruth Christie from the Turkish translation from Kurdish by Muhsin Kizilkaya*

More articles, *Index* archive material and links ➪ www.indexoncensorship.org/turkey
Put your opinion online ➪ www.indexoncensorship.org/comment
Email the Index Index editor for Turkey ➪ gill.n@indexonline.org

# Psychiatry and censorship in the USSR, 1950s–1980s

*The publication of* Censorship: A World Encyclopedia, *edited by Derek Jones, (Fitzroy Dearborn 2001, price £265; £200 to readers of* Index*) marks a first in UK publishing. It is the first ever comprehensive historical survey of censorship worldwide. The culmination of many years' work, with a foreword by Doris Lessing and individual entries written by experts (including staff and contributors to* Index on Censorship*), the* Encyclopedia *is an invaluable – and monumental – work of reference, all four volumes. It is well researched and edited and includes further reading lists after each entry. Below, we publish an edited excerpt in keeping with the theme of this issue* *(see p92).* ***JVH***

'The regular use of psychiatry to repress dissenters in the Soviet Union was inaugurated in the last years of Nikita Khrushchev's regime (1953–64). In the 1970s, under his successor Leonid Brezhnev, the system of 'special' psychiatric hospitals became one of the standard tools to silence voices of criticism, although, as we shall see, ordinary psychiatric hospitals were also used for the same purpose. As the testimony of victims became known in the West, the practice was officially condemned by international groups, such as the World Psychiatric Association in 1977. Mikhail Gorbachev's policies of *glasnost* (openness) and *perestroika* (restructuring) led to repeal of the laws defining 'anti-Soviet' crimes and the release of the inmates of the 'special' hospitals in 1988 and 1989. None the less, even under Yeltsin in the 1990s, the Russian government did not remove the official stigma in the documents of survivors, nor did professional medical associations in Russia admit the extent of earlier misuse.

Khrushchev released political prisoners from Stalinist camps and began investigating the sporadic use of psychiatry to repress and censor. By 1959, however, the tide of reform had receded. Now Khrushchev blamed 'mental disorders' for the resurgence of crime in Soviet society. Forensic psychiatrists, especially those in the Serbsky Institute in Moscow, took the cue to extend the diagnosis of schizophrenia to dissidents, who were said to be suffering from 'reformist delusions'. New facilities were added to the special psychiatric hospital system under the Ministry of Interior (the MVD).

Controls were tightened further in the Brezhnev era (1964–83). For example, the noted biologist Zhores Medvedev was forcibly taken to the Kaluga mental hospital because he had advocated free exchanges with western colleagues. Medvedev was released on 17 June 1970, after only 19 days, as a result of a campaign waged by his twin brother Roy among leading Soviet scientists.

The spurious nature of diagnoses by expert psychiatric witnesses in criminal cases became evident to western observers in the spring of 1971, when Vladimir Bukovsky circulated dossiers on six non-violent protesters: General Petr Grigorenko, Viktor Fainberg, Natalya Gorbanevskaia, Vladimir Borisov, Viktor Kuznetsov and Ivan Iakhimovich. Grigorenko had capped a distinguished military career by becoming a leader of the Democratic Movement in Moscow. A psychiatric commission of the Serbsky Institute had diagnosed him in April 1964 as having a 'paranoid development of the personality, with reformist ideas'. His forced hospitalisation for this 'condition' prevented him from making an articulate defence at a public trial. Grigorenko resumed his campaigns upon his release, and was rearrested while defending the Tatar movement in Tashkent. The local medical commission found him sane, but it was overruled by the Serbsky doctors, who ordered Grigorenko to undergo a new round of compulsory treatment in November 1969 for 'reformist ideas, of which he is unshakeably convinced'.

Fainberg, a philologist, and Gorbanevskaia, a poet, were both arrested for demonstrating against the Warsaw Pact invasion of Czechoslovakia in August 1968. Both had records of emotional difficulties in their teenage years, which allowed Serbsky doctors to claim that they were justified in diagnosing them as having *vialotekushchaia* (sluggish) schizophrenia. This clinical term had been coined by Dr Andrei Snezhnevsky, a senior figure among forensic psychiatrists, to account for the 'outwardly correct behaviour' of persons who became dissidents, as shown by their 'stubbornness and inflexibility of convictions'.

At its height, this misuse of psychiatry affected about a thousand inmates of special hospitals. It was frightening to many other potential dissidents, not as mass terror but as a deterrent. Virtually all such patients were released under Gorbachev, and a new mental health law enacted in Russia in 1992 brought 'coercive psychiatry within the rule of law' and established 'safeguards against violations of human rights', according to Richard J Bonnie, law professor at the University of Virginia. Yet no official apology has ever been issued for past abuses, and the survivors are still listed on police rosters as ex-patients.

***Harvey Fireside***

**RIMAS ZAKAREVICIUS, VIRGINIJUS SAVUKYNAS**

# Language without a people

**With a population of under 300, the Karaims of Lithuania are scarcely able to sustain their language**

Trakai, the ancient capital of Lithuania, 30km west of Lithuania's present capital, Vilnius, is one of the country's most important historical and cultural landmarks. Its impressive Gothic castle, on an island in the middle of Lake Galve at the end of a picturesque peninsula, is usually thronged by visitors from home and abroad. The only way of approaching the castle is via Karaims Street, named after the Turkic nation who, along with other Turkic people like the Tatars, have lived in Lithuania for some six centuries. Today, they number no more than 25,000 worldwide, but have managed to retain a clear sense of identity, as well as their language and religion, the latter based on Judaism.

The Karaims are an interesting ethnic and linguistic phenomenon whose roots reach back to the early Middle Ages. Most now live in Israel, Ukraine, Lithuania, Poland and the USA, with smaller communities in France, the UK, Switzerland, Egypt and Turkey.

There have many attempts, on the whole unsatisfactory, to explain the origins of the Karaims by reference to their name. This is derived from the Hebrew stem *kara* 'to read'; in Hebrew, *karai*, plural *karaim*, means literally 'reading', 'acknowledging only the authority of the reading of the Old Testament'. While this may explain the link between language and religion, it does not illuminate the origins of the Karaims.

Ethnically and linguistically, contemporary Lithuanian, as well as all other eastern European Karaims, are the descendants of the oldest Turkic tribes, the Kipchaks, a term first used in the historical chronicles of Central Asia in the first millennium BC. In European literature they are also referred to as Kumans or Polovtsy. In the fifth century BC, the Kipchaks were living in western Mongolia; in the third century BC, they were conquered by the Huns. From around 600–800 CE, when the first nomadic Turkic empires were founded, the Kipchaks' fate is closely tied to the migrations of the Central Asian tribes.

In the course of the tenth century, having crossed the Volga and settled in the steppes around the Black Sea and in the northern Caucasus, the Kipchaks began to play an important role in eastern Europe. Between the eleventh and fifteenth centuries, they occupied vast territories from the west of Tian-Shan to the Danube. They formed a tribal union ruled collectively by tribal leaders – khans. One of the most powerful of these was the khanate of Khasar, famed for its religious tolerance, unusual at the time. It was as a result of this that Karaim missionaries reached the Khasar as early as the eighth century and converted some Turkic tribes, among them the Kipchaks.

Karaims believe in the one true God and accept the Old Testament, the *Tanach*, as the word of God and the sole religious authority. They recognise no other scriptures, neither the New Testament, the Talmud or any later accretions to the divine word. The most widely accepted theory on Karaimism claims that it started as a reform movement among Mesopotamian Jews in the eighth century during the reign of Caliph Abu-Jafar-Abdullah al-Mansur. Long before the Protestant Reformation in Europe espoused the same idea, Anan ben David Hanassi, the leader of the Mesopotamian movement, was preaching against the existing orthodox rabbinical tradition that left control of biblical exegesis in the hands of the rabbis, and urging Jews to return to the written word of the Scripture and to form their own, personal interpretation of the Word. Ben David's dictum, 'Search ye well in the Scripture and do not rely even on my opinion,' sums up the essence of his reform movement. Islam, the dominant religion in the region, also influenced Karaism and, though the new religion never became a mass movement among Jews in the Near East, it gradually grew in size as it attracted followers from different ethnic groups, including a number of Turkic tribes on the shores of the Black Sea. Their common religion and language united the tribes into a distinct nation, defined by the name of its religion. The Karaim nation of Crimea, Galich-Lutsk, Poland and Lithuania have a shared ethnic origin, religion and language, the latter marked only by differences in regional dialects.

*Trakai castle, 1997: Karaims celebrate 600 years of life in Lithuania. Credit: Karaim Culture Community Archive*

Much Karaim poetry is concerned with love of Trakai, the 'dear beauty' of the former capital and nostalgia for the romantic past. Even love poems, such as this eighteenth-century verse, frequently begin with fond words to Trakai castle, the lake and its island:

THERE IS MY LONGING

How come this city is called a green island?
Its stones are as hard as rock.
And the castle by the lake is an eternal silence . . .
My beloved, how I should like to spend my nights
Beside the lake with you.

*ANDA KIUSIANCHLIARIM*

*Shahary Trochnun otrach tiuviul-mie,*
*Kalasy chiuvriadian suvdan tiuviul-mie,*
*Kiermian karshydan tashtan tiuviul-mie.*
*Ol shatyr, ol chiebriar kiundia, siuviarim,*
*Ingirdia oltuma anda kiusianchliarim.*

Johoshafat Kaplanovski

This popular song, written by a Trakai-born poet a century later, expresses similar sentiments:

DREAMS

Early in the morning I left the house.
Bored of strolling round the streets,
I came to the shore
To the foot of a great mountain.
To whom could I tell my thoughts,
If not to the waves
Driven by the winds
To faraway lands.

*SAHYSHLAR*

*Ertia turup uvdian chychtym,*
*Biezdi kiezmia oramlarda,*
*Suv kyryjha diejin kieldim*
*Bijik tavnyn tabanynda.*
*Kimgia ajtym sahyshymny,*
*Tiuviul-mia bu tolhunlarha,*
*Nie jel siuriadir alarny*
*Astry jyrach kyryjlarha.*

Moisej Pilecki

*Translated by Rimas Zakarevicius*

Religion is the only link between the Karaims of eastern Europe and those in other parts of the world, who call themselves Karaim Jews and are also known as Karaites. A sense of community between Lithuanian Karaims and fellow-believers in Israel, Turkey, the USA and elsewhere is much weaker than, say, among Roman Catholics; Karaims of Turkic origin stress the independent national character of their community.

According to tradition, the history of the Karaims has been intimately linked with Lithuania since 1397–98, when Vytautas, Grand Duke of Lithuania, settled hundreds of Crimean Karaims – and even more Tatars – in the Grand Duchy following an expedition against the Golden Horde. Subsequently, as the dukes developed their territories, many more Karaims were drafted in to populate its more remote and inhospitable desert areas, particularly around Trakai. Initially employed as soldiers and settlers, they eventually became important contributors to the economic development of the Duchy, active as merchants, artisans, doctors and scientists.

And loyal subjects of its rulers. In return for this, the Karaim were granted privileges of citizenship and considerable autonomy under an elected leader, *vaitas*, and spiritual board based in Trakai; the first extant document listing Karaim rights to self-government dates from 1441. It was largely this that, despite their integration into the Lithuanian state, enabled the Karaim to avoid assimilation and to retain independent contact with the Karaim of Crimea and Galich-Lutsk.

With notable exceptions – when they became the victims of anti-Semitic pogroms at the end of the fifteeenth century, and again in the nineteenth when Lithuania came under Czarist Russia – relations between the Karaim and the local population were peaceful; their historic links with Duke Vytautas, seen by most Lithuanians as their greatest ruler and instigator of their country's 'golden age', along with a secluded lifestyle in which they practised their religion, and preserved their language, literature and folk tradition, largely contributed to their survival. They were not, in any case, ever sufficiently large in number – at most between 3,000 and 5,000 – to constitute any threat to the locals, and were concentrated around the Trakai area. In World War II, they were protected from the occupying Nazis by stressing their Turkic origins rather than the 'jewishness' of their religious practices. Eventually, they abandoned the use of Hebrew in favour of Karaim for liturgical purposes, a step in the preservation of the language.

The oldest written record of the language is a hymn translation, published in Venice in 1528. The first Karaim periodicals appeared at the beginning of the twentieth century and continued to witness to the vitality of the language and community until Soviet times; at this point, Karaim self-expression and their consciousness as part of a separate nation fell foul of Soviet 'internationalism' as Lithuania was absorbed into the USSR. All separate minority cultural and religious activities throughout the empire ceased, until the end of Soviet power and the restoration of Lithuanian independence in 1991. Renewed political and cultural freedom saw the revival of Karaim consciousness, language and religion. Lithuanian Karaims and their relatives in Poland (c150 people) spearheaded a cultural and religious revival among the approximate 2,000 Karaims in the Crimea and other parts of the former USSR.

In 1992, the Religious Community of Lithuanian Karaims recovered the ancient community rights they had been granted in the fourteenth century, and was once again recognised as a traditional religious community of Lithuania. In 1997, they officially celebrated the six-hundredth anniversary of the settlement of the Karaims and Tatars in Lithuania.

Today, despite favourable conditions for the expression of their national identity and language, the Karaim face a serious problem: the apparently inexorable decline in population. At the beginning of 1997, a Lithuanian-wide survey identified only 257 Karaims, 32 of them children under 16 years old. The number of fluent speakers of Karaim is low: only 50 or so people in Lithuania still use the language in everyday conversation. Though still functioning in private life and liturgical ceremonies and being learned by the younger generation in Sunday schools, the language of the Karaims, failing a significant rise in population, is doomed to extinction along with the identity of its speakers. ❑

***Rimas Zakarevicius*** *is an historian.* ***Virginijus Savukynas*** *is a cultural anthropologist and editor of the web magazine* Omnitel Laikas

WWW
Picture feature ⇨ www.indexoncensorship.org/endangeredlanguage
Put your opinion online ⇨ www.indexoncensorship.org/comment
Email the editor ⇨ editor@indexonline.org

JOHN MUTENDA MBILI

# Summons to enter

**Who killed Fr John Kaiser? A Kenyan gumshoe makes his own enquiries**

There are 54 banks in Kenya plus another 100 or so other financial institutions and building societies. Eighty per cent of all banking business is dominated by four major banks. We 'investigators' serve the banking and insurance sectors as debt collectors and enforcers of bad debts.

Borrowers come in all shapes and sizes and their reasons for default are as various. Genuine borrowers go to the bank to get the capital to invest in business or assets. Most of these default for one of two reasons: lack of business skills or conditions beyond their control. A second category of borrowers, political borrowers, will always default. They borrow on behalf of their political masters to run their campaigns or execute their orders. They have no intention of paying a debt which is theirs in name only and they are, naturally, protected by those in power on whose behalf they have contracted the loan.

Nevertheless, some effort must be made to recover the money and this is where we come in. Usually loans are secured with tangible assets, but political loans are either inadequately secured or not at all. Even when the loans are secured, the country's interest rates are abnormally high so that even after several months of irregular repayments, borrowers find themselves still owing a large part of the original debt.

The government is instrumental in determining these high levels of interest. It collects the taxes that are stolen and squandered by those in office. Finding themselves unable to provide public services, they head off to London or Washington, arms outstretched, on begging sprees. Often they manage to convince the lenders – commonly known here as donors – that they have important projects for the good of the people. The lenders are usually hoodwinked into parting with their money.

Every once in a while, when the salaries of soldiers and civil servants need to be paid, the government finds itself in a tight spot and turns to the domestic market via treasury bills and bonds targeted at the banks. Seduced by the high rates of interest, the banks cannot resist the temptation; these inflated rates are then handed on to the ordinary borrower.

There are standard procedures for recovering loans, starting with a letter from the bank. We are only brought in when all else has failed to elicit the money. Our assignments are governed by the creditors' or their lawyers' need to know the whereabouts, means and assets of the debtor so that these can be attached once court orders have been secured. It takes between three days and a month to find our quarry and the assignments take us into every corner of this country. We take on the mighty and the weak; we look for them in their homes, at their places of business and even the places where they take their recreation.

When they discover that we are after them, they systematically avoid us by hiding, running away and – for those who are daring enough – by using their positions to resist. In the latter case, they may threaten or use force to eject us from their premises. One cabinet minister, still serving in the government, had me forcibly thrown out of his office by his bodyguards when I attempted to serve him with a Summons To Enter Appearance. In another instance, another minister ordered armed guards to frogmarch me from his home compound to the gate with the order: 'and in case of any resistance, shoot him dead!' In yet another case, a serving minister ordered the security detail at his home to warn me that, should I ever go back there, they would shoot me on sight. Another tycoon had my assistant eat a pair of summonses and a copy of a plaint – five A4 sheets in all – with a bottle of Fanta he had bought especially for the purpose. Incidents of this kind are legion in our daily lives.

As an investigator, I have been threatened with death countless times and been forcibly ejected from the debtors' premises. This does not impair our efforts nor our open-mindedness. But it does, however, make me think of the murder of Father John Kaiser, 68, and the way it has been investigated.

Kaiser's body was discovered under a tree about 3km from Naivasha on the highway towards Nakuru in the early hours of 24 August 2000 by traffic police. The back of the priest's head had been blown off by a shotgun. Though not specialised homicide investigators, the police's initial impression was that he had been murdered. They did a remarkable job of preserving the scene for their colleagues from the Nakuru CID who, before conducting any investigation, concluded on the spot that the priest had committed suicide. This sparked a roar of outrage, led by the Catholic Church and with the support of other denominations countrywide. Fr Kaiser was no ordinary man. After 36 years working as a priest in Kenya, he had become a national icon, a symbol of social justice and a fighter for human rights.

His work was first reported by the national media when he took personal charge of the Maela Camp where internal refugees, displaced by tribal clashes between the Kisii and the Maasai that Kaiser claimed had been provoked by the government, had been abandoned. He set up a temporary camp, complete with schools for the children and a dispensary, and solicited from well-wishers locally and overseas. Though the clashes were publicly condemned by the government, government forces were terrorising the victims, killing them rather than rescuing them. Fr Kaiser exposed the atrocities outside Kenya. As a missionary, his testimony was credible and he became the official spokesman for the refugees. Which did not please the architects of the clashes.

He was threatened and verbally abused. When his work permit expired, the government refused to renew it; only the intervention of the US Ambassador succeeded in getting the document renewed. At the time, the government was keen to convince the world it had no hand in the tribal clashes, particularly the donors who run this country from the comfort of Washington, hence the setting up of the Akiwumi Commission of Enquiry.

Fr Kaiser's testimony was the single most revealing account of the extent to which the government and its officials were involved. While ordinary citizens enjoyed reading the revelations in the daily papers, it was highly embarrassing to the establishment. Indeed, some of us who know the ruthlessness of the local mafia that run this country were surprised that he was not killed before finishing his lengthy testimony that lasted over several days.

Kaiser's concerns did not stop with the refugees. In the course of his ministry, he had learned that his parish was located in an area of high illiteracy. A major concern was the forced early marriages of Maasai girls and the high rate of rape among schoolgirls. He had evidence that a

*Fr John Kaiser in happier times. Credit: Courtesy Mary Mahoney*

minister, a member of his parish, had raped and impregnated a 13-year-old girl. While Kenya is by no means a normal society – it is ruled by thugs called 'ministers' who rape and steal, are corrupt and scandalous yet remain in office – Kaiser took the steps any normal person would have taken in a normal society: he enlisted the assistance of the Kenyan Federation of Women Lawyers (FIDA) to prosecute the offender.

It happened that the minister in question was in charge of the police. As a result, the FIDA lawyers who were handling the case received death threats. The girl's parents were also intimidated until they had persuaded the girl to withdraw her complaint. Where did all this leave Fr Kaiser?

Once the case had been withdrawn, it was time to deal with Kaiser. He received verbal threats, his house was attacked, he was tailed and his movements recorded. He mentioned his fears about this psychological harassment to various people, among them lawyer Muite, FIDA lawyer Kuria, Davinder Lamba, Fr Tom Keane, Romulus Ochieng, Fr Mwangi, his bishop Colin Davies, Brother Martin, Fr Boyle, Sister Nuala and the Papal Nuncio. Despite all the evidence of a man under siege, and the fact that the local CID had already concluded that Fr Kaiser had committed suicide, why did it take the US Federal Bureau of Investigations seven months and vast public resources to come to the same conclusion?

Suppose we agree with the FBI that Fr Kaiser committed suicide, that he was suffering from manic depression, that they did their best to gather all available evidence before coming to this conclusion, does it mean that the condition was not treatable? Their report, released on 20 April this year, says they reconstructed the victim's last 96 hours of life – but admits they do not know what he was doing in the last three of those hours. Nor do they know what happened at the scene of the crime between 3am and 5.45am when the first cops arrived; nor what happened to the scene from the time it was taken over by the police until they arrived.

The FBI do not seem to have been aware that the Kenyan police is politically manipulable and, therefore, permanently compromised to the extent of being willing to arrange the scene for a premeditated outcome. If indeed this was the case, the FBI seems to have become partners of the Kenyan police in perpetuating the official cover-up. Is it a wonder then that no one in this country seems to take the FBI report seriously?

Following the release of the official report, the bishops met for a week to examine it fully and the Catholic Church has now issued its official response. They raise several pertinent questions challenging the FBI's professionalism. They note that the report does not include the ballistics report. The pathologist's report indicated that there were bloody fingerprints in the deceased's pocket; how did the fingerprints get there, the bishops asked? Or did he put his hand in his pocket after he blew off his head? They have rejected the report as a farce and demanded a full public inquest into the death.

The victim's sister, Carolita Mahoney, who was in Nairobi when the FBI report was released, does not believe her brother committed suicide. She has vowed to institute fresh investigations to establish the truth. 'They assassinated his body, and now they want to assassinate his character,' she was quoted as saying. Bishop Davies said that the situation in Kaiser's parish, Llogorian, was more threatening now than a year ago: 'It's a very nasty situation in Kenya,' he said in a visit to the US in August this year. 'The FBI has given a green light to more murders in Kenya. The Moi government now knows it can get away with murdering an American. They will have no fear of anything anymore.' He added: 'It was a good attempt to silence any further questions about [John's] death, but we're not being quiet – and we won't be.'

The independent Anti-Corruption Network Secretariat has dismissed suicide as the cause of Fr Kaiser's death on the grounds that the FBI may have been misled by the Kenyan police. They have promised to institute fresh investigations. Meanwhile, Kenyan bishops have requested a formal inquest into Kaiser's death. ❑

***John Mutenda Mbili** is a private investigator and debt collector in Kenya*

Support for

# INDEX ON CENSORSHIP

It is the generosity of our friends and supporters that makes *Index on Censorship*'s work possible. *Index* remains the only international publication devoted to the promotion and protection of that basic, yet still abused, human right – freedom of expression.

Your support is needed more than ever now as *Index* and the Writers and Scholars Educational Trust continue to grow and develop new projects. Donations will enable us to expand our website, which will make access to *Index*'s stories and communication between free speech activists and supporters even easier, and will help directly in our education programme. This will see *Index*, for the first time, fostering a better understanding of censorship and anti-censorship issues in schools.

Please help *Index* speak out.

The Trustees and Directors would like to thank the many individuals and organisations who support *Index on Censorship* and the Writers and Scholars Educational Trust, including:

The Ajahma Charitable Trust
G E Anwar
The Arts Council of England
The Bromley Trust
Clarins UK Ltd
The John S Cohen Foundation
DANIDA
Demokratifonden
The European Commission
Nicole Farhi
The Ford Foundation
The Goldberg Family Trust
The Golsoncot Foundation
The Calouste Gulbenkian Foundation
G A Hosking
Institusjonen Fritt Ord
The J M Kaplan Fund
Dr and Mrs J Matthews
Mr Raymond Modiz
The Norwegian Non-fiction Writers and Translators Association
Offshore
The Open Society Institute
The Onaway Trust
Pearson plc Charity Trust
Thomas Pink
Proud Galleries
C A Rodewald Charitable Settlement
Ms Ruth Rogers
The Royal Literary Fund
The Royal Ministry of Foreign Affairs, Norway
Alan and Babette Sainsbury Charitable Trust
Scottish Media Group plc
J Simpson
R C Simpson
Stephen Spender Memorial Fund
Tom Stoppard
Swedish International Development Co-operation Agency
Tiffany & Co
Ms E Twining
United News and Media plc
UNESCO
Mr and Mrs O J Whitley
and other anonymous donors

If you would like more information about *Index on Censorship* or would like to support our work, please contact Hugo Grieve, Development Manager, on (44) 20 7278 2313 or email hugo@indexoncensorship.org